Be Unforgettable

Handbook for Students and Teachers

Chakradhar Dixit

Table of Contents

Introduction

Section I: Articulate your brain to Enhance Memory Recall

Chapter I: Five Unproductive Practices to Stop Immediately

Chapter II: Five Recommended Practices to get Better Memory Retention

Section II: Different Memory Training Techniques

Chapter III: Association and Pegging Method

Chapter IV: Chain Linking Method and Tagging or PNN Method

Chapter V: Mnemonics

Chapter VI: Loci Method or Memory Palace Technique or Roman Room Method

Chapter VII: Number Sound Method or Phonetic Method

Chapter VIII: Mind Maps®

Section II: A New Approach to Old Learning

Chapter IX: Mathematics

Chapter X: Physics

Chapter XI: Chemistry

Chapter XII: Biology

Chapter XIII: Foreign Languages

Chapter XIV: History

Chapter XV: Geography

Introduction

A gentleman was walking through an elephant camp, and he spotted that the elephants weren't being kept in cages or held by use of chains. All that was holding them back from escaping the camp, was a small piece of rope tied to one of their legs.

As the man gazed upon the elephants, he was completely confused as to why the elephants didn't use their strength to break the rope and escape the camp. They could easily have done so, but instead, they didn't try at all.

Curious and wanting to know the answer, he asked a trainer nearby why the elephants were just standing there and never tried to escape.

The trainer replied "When they were young and much smaller, we use the same size rope to tie them and, at that age, it's enough to hold them. As they grow up, they are conditioned to believe they cannot break away. They believe the rope can still hold them, so they never try to break free".

The only reason that the elephants weren't breaking free and escaping from the camp was that over time they adopted the belief that it wasn't possible.

Here elephant is your brain and rope are your limited beliefs which have hold you till date to explore what you can achieve. This book is going to be your key for your subconscious brain to break free from that rope and will give you the life you desired.

I have learned certain memory skills from by mentors and I have applied myself and written one book in a months' time only for beginners and now I am coming up with this new book which will cover most of the topics which have been taught all over the world. This is not only a guide for students but also to all their mentors/teachers/trainers i.e., how to teach in fun and easy way for maximum comprehension.

Before beginning our fun ride, I want to thank you first for starting your transformation journey of new learning with me. I also want to thank all my mentors who have given me the most crucial self-confidence that I can do it. I want to thank **Mr. Jim Kwik** for a short but life changing lessons given through his masterclass. I also want to say thanks to **Mr. Vishen Lakhiani** to show me the light after the tunnel concept and I have continued the journey without stopping. I cannot forget the contribution of concepts learned from **Ms. Aditi Singhal** which motivated me to learn more and more concepts related to academics and in turn share it with my younger friends who are struggling with their academics. I even would not have started this journey If I have not attended **Rohini Mundra**'s Extraordinary Masterclass. She has given me the belief that anyone can reach to the 1%club for that you just need perseverance, nothing else and you keep putting your one foot in front of the other, and then one day you look back and you've climbed the mountain. Finally, I want to thank my parents and my entire family for believing in me even in tried times.

I will tell you first, some basic steps to bring few changes in study routines and then I will share the concepts which I believe will completely change your perspective about learning any academic topic i.e., you will start adoring your learning process as you will be more involved and your comprehension will also be improved. But need one promise before starting that you will read the whole book at least once as I have told many concepts and you never know which concept will work for you. I have tried to incorporate maximum course ideas from across the world but if something is missing, I will love to listen it from you so that same can be included in further series of books which is going to be at the experts' level. I have tried to explain the concepts using stories because stories are more effective and accessible than arguments and explanations. Most students seem to understand stories better than logical reasoning and arguments. Stories add to cultural literacy, invigorate presentations and brings interest. Stories also connects well and get students thinking and asking questions. To tell you in short, stories let us think the way our ancestors were thinking at the time of invention in science or coding in computers or problem solving in math and make it more personal to us.

Section I: Articulate your brain to Enhance Memory Recall

Gautama Buddha

We are shaped by our thoughts; we become what we think. When the mind is pure, joy follows like a shadow that never leaves.

Chapter I

Five Unproductive Practices to Stop Immediately

We will first start with few things which you should stop doing immediately. Yes, forget these things from today for better, brighter and smarter brains. This book is about memory training and I am telling you to forget in very first chapter, as there are some specific reasons behind it. As the saying goes "empty your cup so that it can be filled with better and healthier alternative ", this is your process of emptying your cup and explore your superhuman capabilities from very beginning.

You have to forget these five things immediately and those things are: -

Multi-Tasking:

What you call multi-tasking is really task-switching, (says Guy Winch, PhD, author of the Emotional First Aid: Practical strategies for Treating Failure, Rejection, Guilt and Other Everyday Phycological Injuries). Switching between tasks affects your productivity to great extent because your attention is not completely focused on any one task hence your mind does not allow you to balance any one task. In fact, multitasking will probably take you longer to finish two tasks when you are jumping back and forth than it would to each one separately. Each task requires a definite mindset, and once you have channelized your mind for any one task, you should stay there and finish. Multitasking causes stress both physically as well as mentally as you have not finished both of your tasks. For example, if you do poorly on an exam because you studied while watching a baseball game on TV, that can certainly trigger a lot of stress—even self-esteem issues and depression.

Multitasking occupies lot of pace in our "working memory," or temporary brain storage, in layman's terms. And when working memory's all used up, it can limit our ability to think creatively. So, you're hurting your brain capability more than helping yourself by multitasking and you should stop it immediately.

Rote Memorization:

Most of us have grown up with the ability to rote memorization since childhood i.e., when we started memorizing the alphabet, numbers, times tables and formulae. We are simply acquiring the knowledge given, which is causing more damage to our imagination than gain. We never think that there can be more logical answer to the given question. We have taught our children that bookish knowledge is paramount and you do not need to put much brain to get external knowledge. Rote memorization somewhat unable to prepare the students for outside world, also impairs their artistic thoughts and affects critical thinking process. Rote learning makes studying boring, uninteresting and completely unengaging. Unless the question is a mathematical one, there may be more than one answer which is correct, but the rote learner will never develop the ability to explore the options that lead to the different answer. Rote memorization never let them learn to question and explore. These people develop their listening and writing skills but not their thinking and questioning skills. Rote memorization relates more with short-term memory. Apart from certain exceptions like the times tables and period table values, most rote learning is for those who want the knowledge for a certain purpose. For example, a student might learn the Periodic table for an exam but will almost immediately forget the instances in which the application of acquired knowledge might be used. Rote can be considered a "quick-fix" solution to gaining knowledge. The teacher is expected to tell the students the complete process of solving any problem and inspire them to find out how the answer was drawn. The students, on the other hand, will accept the teacher's version of the answer without questioning the method. This leads to the superficial knowledge of any given subject.

If a particular question might require solution from a different angle, the student will never be able to answer it because they have not been taught to. So, Students who learn with creative or pictorial learning are able to solve the problem better than those who learn by rote. You have to simply understand the concept. Once you understand the concept it will never be required to remember. There are different ways of learning except rote memorization. You can try different techniques described in this book and you may never to memorize again using rote memorization.

Cramming:

Cramming usually happens when students put off studying till the very last time. Students spend hours memorizing as much of the material as possible in a short period of time, like the night before the test. They may stay up all night, and they think they are doing lot of work. This type of studying may become repetitive, or become the only way that a student knows how to study. Cramming is one of the least effective ways to learn a subject. Research has found that many students cannot recall much information after a cram session. They have trained their mind to recite the material without developing a deeper understanding. This weakens the learning process. The student has lost a learning opportunity by simply momentarily memorizing everything for a decent grade. A big reason why cramming doesn't work is that it vividly increases a student's stress levels. This stress has a negative effect on their ability to concentrate, making preparing for a test even more difficult. Cramming for tests is a short-term solution. The information students spend the night cramming is stored in the short-term memory where it is accessible in the short-term (like for a test the next day). However, it doesn't create long-term neural connections to the material, or develop profound grip on the subject. Students might get a good score on the test, but that knowledge will be elapsed quickly – usually as soon as the following days. Rather a technique known as "spaced repetition" is much more effective for short-term recall and long-term retention.

Spaced Repetition is when students review the material over a long period of time. This gives their minds time to form connections between the concepts and notions. This knowledge can be built upon and easily recalled later. I will share the techniques with you for effortless space repetition and effective recall later in this book.

Unplanned Study Schedule:

Studying at any level requires decent time management, and if you find yourself struggling to meet deadlines, or you feel overwhelmed with work, or you frequently end up having to stay up late into the night to finish off a piece of homework, this is a sign that you need to work on your time management skills. This means becoming more organized, keeping a list of what needs to be done and by when, and getting started on homework as soon as you're set it, rather than putting it off. It also means being more disciplined with your routine: getting up earlier, planning out your day, and making maximum productive use of the time you allocate to each of your subjects. If you're stuck at any point, take a little time out, clear your head and adjust your way of thinking about your studies. One such technique is "Pomodoro Technique" which will improve your productivity multi fold.

The Pomodoro Technique was developed in the late 1980s by then university student Francesco Cirillo. Cirillo was struggling to focus on his studies and complete assignments. Feeling overwhelmed, he asked himself to commit to just 10 minutes of focused study time. Encouraged by the challenge, he found a tomato (*pomodoro in Italian*) shaped kitchen timer, and the Pomodoro technique was born. The steps involved in Pomodoro techniques are: -

1. Get a to-do list and a timer.
2. Set your timer for 25 minutes, and focus on a single task until the timer rings.
3. When your session ends, mark off one pomodoro and record what you completed.
4. Then enjoy a five-minute break.

5. After four Pomodoro's, take a longer, more soothing 15-30-minute break.

In the event of an unavoidable distraction, take your five-minute break and start again. The rule applies even if you finish your given task before the timer goes off.

Distractions:

Your mind gets distracted when your work is interrupted in flowing state. If you continue to receive notifications on your mobile regarding emails, group chats, and social media updates, how will you be able to concentrate? It is easy to blame technology but this is more of self-inflicted. In the moment, you convince your mind that this is only for few moments and your mind also supports you by telling — "Other things wait but this email cannot," or "It will take hardly a minute to check my Facebook page or any twitter update." These small interruptions take up a lot of your precious time. It also takes time and energy to refocus your attention. Indulging the impulse to check Facebook "just for a minute" can turn into 20 minutes of trying to get back on task. Some solutions for keeping distractions far away are: -

i. **Take a deep breath when you're about to get distracted-** Close your eyes. Breathe in for 5 seconds and exhale. Repeat this action minimum 4-5 times. Repeat this process if your still unable to focus.
ii. **Turn off your Internet access-**Internet has become integral part of your computer. It is repository of knowledge and also a means of entertainment, it is up to you how to use it. Turn off your Internet access

before you begin your study session. If you need to access certain online resources, then download all of the necessary information at the start of your session before you turn off your Internet access. By turning off your Internet access when it's time to focus, you're utilizing the power of the Internet effectively.

iii. **Put your phone on silent mode and place it at the other end of the room-** At the start of your study session—you put your phone on silent mode and place it as far as you can. Preferably, you should place it at the other end of the room. This way, you can avoid interference by phone calls or text messages while you're studying. You can always check your phone every 30 or 45 minutes when you take a break.

iv. **Take one step at a time on difficult projects-** Difficult tasks are more likely to distractions because they're pathogenic, for obvious reasons. And when we feel pain, our natural tendency is to call for bliss (examples: Facebook and YouTube), which is nothing but distraction. So, if a task is cumbersome, proceed gradually. If you attempt too many things in one session, you're likely to fall to distractions.

v. **Schedule Distractions-** After some efficacious continuous sessions, go and satisfy your urge to surf the internet, text, and chat. Plan them all together and finish it, preferably, in a 30-odd minute session. Once done, start your next chain of undistracted sessions.

In our ever more connected world of smartphones, tablets, laptops and high-speed Internet, distractions are inevitable. You'll need to make a cautious and committed effort to stay on task. Avoiding interruptions is not easy, but is feasible. Like any skill, this too requires practice and discipline to achieve expertise. Take control of your mind and get rid of all the wrong study practices from this very moment.

Chapter II

Five Recommended Practices to get Better Memory Retention

As if we have understood what we should not do, it is time to understand what we have to include in our learning process to make it more emphatic. I will tell you five things that will make your learning process a fun experience. Those five things are: -

Creative Note- taking:

There are many ways to take notes. It's helpful to try out different methods and determine which work best for you in different situations. Whether you are learning online or in person, the physical act of writing can help you remember better than just listening or reading. Research shows that taking notes by hand is more effective than typing on a laptop. I am going to share two such techniques one is Cornell Note taking system and another technique is Mind Maps® which is devised by great Tony Buzan.

"Cornell Note Taking" is a system for taking, organizing and reviewing notes (Cornell Note Taking has been devised by Prof. Walter Pauk of Cornell University in the 1950s). It requires slight preparation which makes it perfect for note taking in class. The page will be separated into Four different sections: Two columns, one area at the bottom of the page, and one smaller area at the top of the page. An intelligible process of note taking you can say. All actual notes from the lecture go into the main note-taking column. The smaller column on the left side is for questions about the notes that can be answered when reviewing and keywords or comments that make the whole reviewing and exam preparation process easier. When reviewing the notes, a brief summary of every page should be written into the section at the bottom. Besides being a very effectual way of taking great notes in class, Cornell note taking is perfect for exam preparation. The system encourages you to reflect on your notes by actively summarizing them in their own words. Often, this can already be enough to remember study notes and to successfully pass an exam. When reviewing your notes, it's also convenient to restructure objects on the page, for example, to add a solution to an answer on the side to the notes.

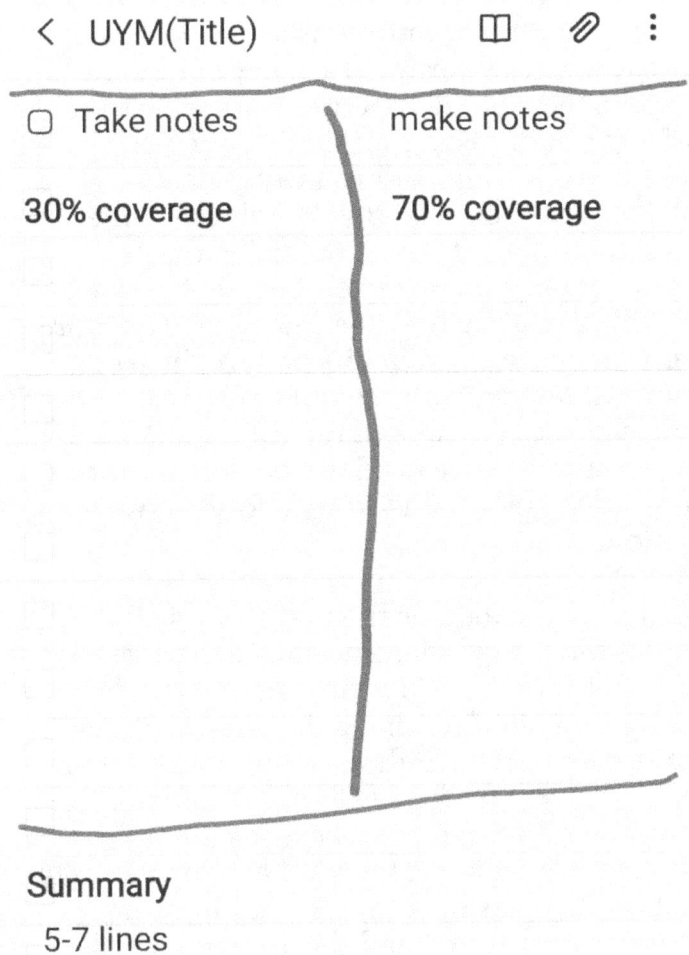

(An example of Cornwell Note-taking method)

Spaced Repetition:

Spaced repetition is an ancient technique for efficient memorization & practice of skills where memorization can be done far more efficiently by spacing out each review, with increasing durations as one learns the item. Because of the greater efficiency of its slow but steady approach, spaced repetition can scale to memorizing hundreds of thousands of items, whereas crammed items are almost instantly gone. For example, let's say you read that "Canberra is the capital of Australia." If you're not using that information at all, you will likely to forget about it after finishing reading or sometime later. The more often you encounter certain bits of info, the less often you'll need to refresh your memory of it. Now question is how to use spaced repetition for effective learning. Before that we need to understand our forgetting curve and that is why we forget any information.

Hermann Ebbinghaus was the first to study the behavior of forgetting scientifically. He performed tests on himself over various time periods, then analyzed the data to find the exact "shape" of the Forgetting Curve. {His findings were published in 1885 in the book (Written in German) "Memory: A Contribution to Experimental Psychology."} According to Ebbinghaus, the level at which we retain information depends on a couple of things: the strength of your memory and the amount of time that's passed since learning. The graph below represents memory retention over the span of a week. The forgetting curve estimated that after two days, **only 25% of information is retained**. Other studies have proven similar results with estimates that **70% of memory is lost within the first 24 hours**.

The forgetting curve

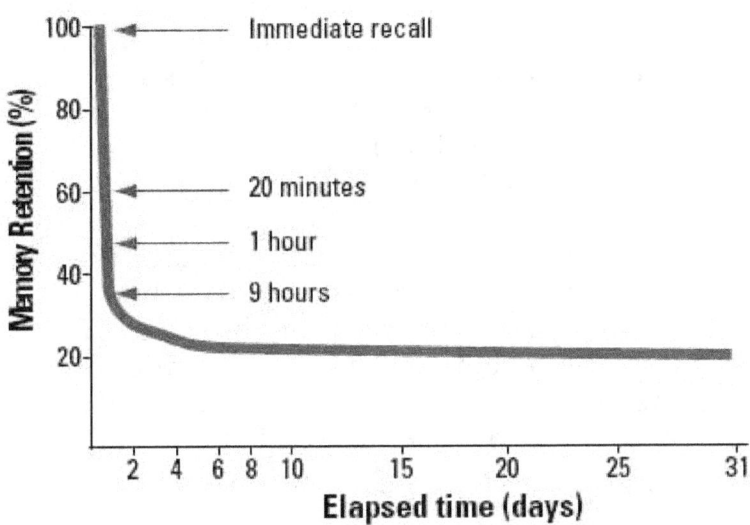

The "forgetting curve" was developed by Hermann Ebbinghaus in 1885. Ebbinghaus memorized a series of nonsense syllables and then tested his memory of them at various periods ranging from 20 minutes to 31 days. This simple but landmark research project was the first to demonstrate that there is an exponential loss of memory unless information is reinforced.

The speed at which we forget any information depends on a variety of factors:

i. The difficulty of the learned material
ii. How easy it is to relate the information with facts, which are already known
iii. Quality of memory representation
iv. The conditions under which the material is learned
v. Whether the student fully rested or stressed

Spaced repetition, is a method of learning that involves breaking down and revisiting information over time, causing long-term retention to drastically improve. The more we practice and the more spaced this repetition becomes, the more likely we are to encode this information into our long-term memory. In essence, the idea behind spaced repetition is that you allow your brain to forget some of the information to ensure that the recalling process is mentally strenuous. The psychology literature suggests that the harder that your brain has to work to retrieve information, the more likely that information will be encoded. What's even more astounding is that evidence suggests that, even within the same study session, spaced repetition can be a more efficient technique in terms of retaining information. Our review schedule should be as given in the chart below i.e.

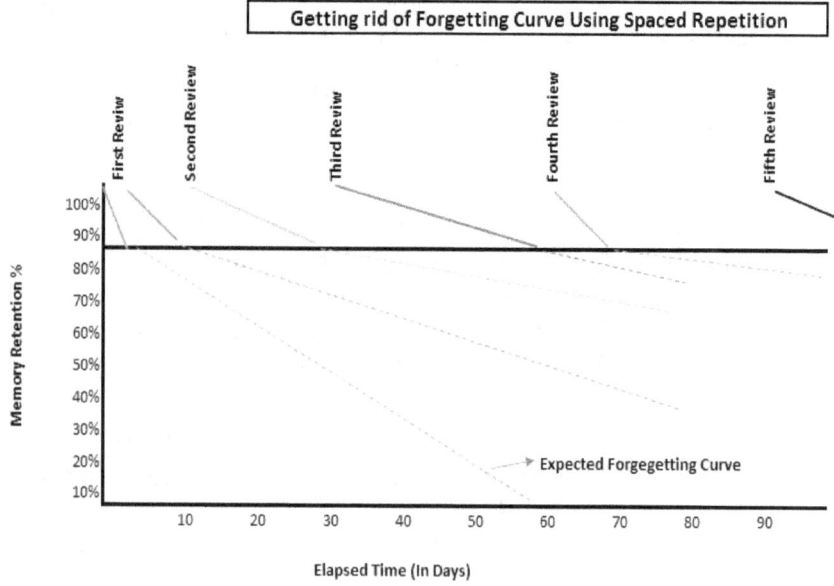

i. First review: immediately after an hour
ii. Second review: 24 hours later

iii. Third review: one week later
iv. Fourth review: one month later

Finally, when you will be reviewing it fifth time i.e., in three months or before your exams, you will find that you have at least 95% or more retention. Also, you're reviewing or revision time will be minimum and you will be very confident as you can recall most of the information. You can also schedule your spaced repetition using your mobile apps also like Ankidroid *(Anki Japanese word for Memorization)*, Quizlet, Memrise and Cram.com o. These apps have spaced repetition algorithm that makes memorization more efficient.

Active Recall:

Active recall is efficient way to transfer any information from short-term memory to long-term memory so that you can get it when you need it. According to research, active recall performs 50% better than the normal studying. In active recall, once you finished referring your study material, close your eyes and recall every bit of the information and then open your eyes and write it down and finally check your comprehension i.e., how much you have retained from your notes. Once you will make this a practice then your comprehensions will go on increasing gradually and you will always be able to recall any information in short span of time. Aristotle once said" Exercise regularly recalling a thing strengthens the memory". Active recall proves more helpful than other techniques for you to remember, how? Active recall is mainly based on re-reading, concept mapping and retrieval.

i. Whenever you put yourself to test and retrieve any information, you're are actually strengthening the connections between your neurons in your brain by creating additional associations between them
ii. When you try to recall any information, you think more and thus you lead your brain towards a streamlined thinking i.e., striking out irrelevant responses and focus on definite answer which you need.
iii. As you ask specific question while recalling those questions becomes your cue for any information when in you require to put in practice like exams or real-life situations

Some of the active recall techniques used by active learners across the world are: -

Flashcards: - Make your unique and different flash cards and try to recall the information with your spaced repetition schedule. You can use mobile apps like Anki, Memrise etc. but keep it in mind that personalized flashcards will always works well for you

Teach it to someone: - Find a friend who also need that information to study it for his exams and share with each other whatever you have recalled and chances are you both will learned in twice.

Create Mock Tests: - Try to create exam like environment while studying and you will able to handle the pressure of real exams

Chunking: - Break the points in small bite size information which you can handle and make your study notes for your spaced repetition schedule

Active recall helps a lot in remembering and learning things that you study in very effective manner. Once you start following the method, learning any new thing will become fun and interesting for you.

Multisensory Learning:

Neuroscience and cognitive psychology have unveiled the extraordinary power of senses in learning process. According to researchers our senses process any information by sight-83%, Hearing- 11%, Smell-3.5%, Touch-1.5% and Taste-1%. So, it is better to bring all our senses in the learning process to get maximum benefits from them. Also, every individual has different learning techniques. Some people learn by watching, someone by listening and someone learns it by their hands-on experience. These three learning techniques are called Visual Learning (Eyes), Auditory Learning (Ears) and Kinesthetic Learning (Touch). Two remaining senses are taste and smell.

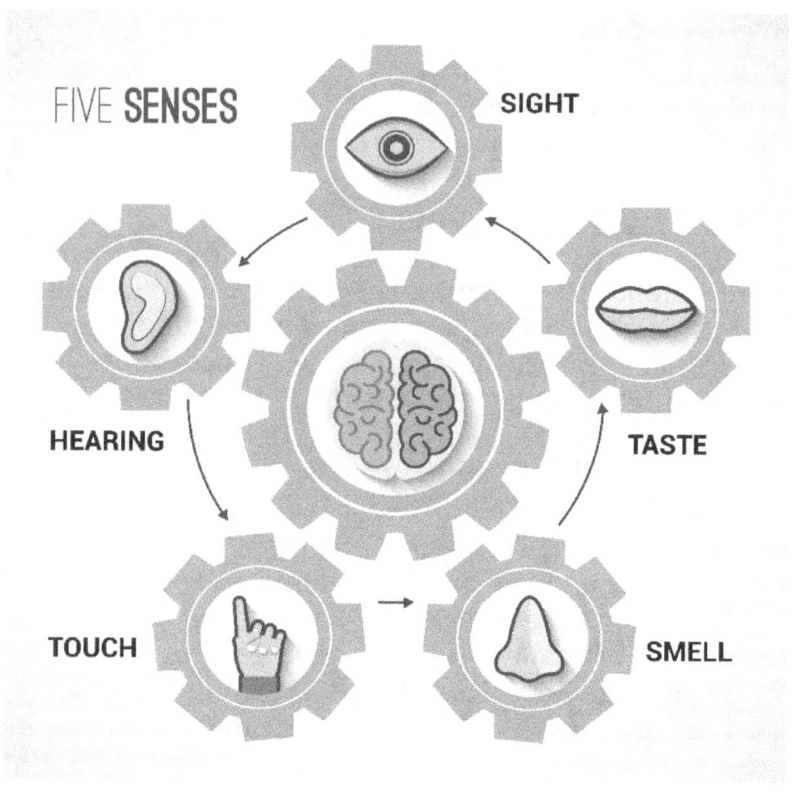

ii

We respond almost immediately to most sensory information. It is important to know how our five senses, synchronized with each other to facilitate better learning. How multisensory learning improve our concentration, alertness, memory, mobilization creativity and communication we will understand one by one. There are many ways to bring our senses in our study and we will see few of them for our better understanding.

Sense of Sight: Visualization works from a human perspective because we respond to and process visual data better than any other type of data. In fact, the human brain processes images 60,000 times faster than text, and 90 percent of information transmitted to the brain is visual. Visuals in the classroom via infographics that use large images can reinforce key concepts, and colors that stimulate mental activity. So, first of all how can you ascertain that you are a visual learner or not? If you like doodling on paper, have good dress sense, and if your friends and colleagues find you creative in terms of designing, photography or you like something that has good sense of orientation and planning. One final tip when you asked a question and your start looking up i.e., start visualizing the picture associated with that question you are definitely a Visual Learner. You can put your learning technique to work, you should start doing following things in your learning process

i. Start taking notes mostly in the form of mind maps, pictures, diagrams or images whichever suit for you
ii. Use different colors mostly bright colors while taking notes or self-study i.e., highlighting important keywords. Colors are known to activate learning like Red color- brings your attention towards learning, yellow- prevents mental activity and blue color- can make you calm if you are feeling overwhelmed
iii. Use number Shape method and number association methods for memorizing lists (These methods will be discussed in detail in later chapter so don't worry for them as of now)

Sense of Hearing: Hearing helps us in learning associations, and it is crucial part of our learning process throughout life. One can identify an auditory learner by asking him any questions and if they are looking sideways i.e., they are trying to recall where they have heard it, they are your auditory learners. If you ask lot of questions in the class, love listening to music while doing your chores and find yourself more often cheering up in school or college games undoubtedly your style is auditory learning. You should utilize it completely by incorporating following habits in your leaning process.

i. Include Baroque music in your learning process, as research suggests that students had almost 92% retention while listening to baroque music or any classical music up to 60 beats per minute (strictly instrumental music without lyrics). Music activates our right brain and as we are listening to soothing music our blood pressure decreases as heart and pulse relax to the beat of the music
ii. Listen 428hz music while meditating, studying or exercising to get maximum benefits out of those activities.

iii. Discuss the topic in group of encouraging colleagues and take feedback
iv. Record yourself while revising your notes and listen to it
v. Use rhyming mnemonics to revise your notes (It will be explained in detail in later chapters)

Sense of Touch: As explained above sense of touch is associated with Kinesthetic Learners who learns better by hands on experience i.e., by practically doing it. These learners are good at sports and dance. They look down whenever they have been asked any question i.e., they try to find out answer by writing it. Teachers can identify them easily as these are the students whom teacher always says to "concentrate, focus or sit properly etc. in the classroom. they are the one who cannot sit idle. But these traits of you can be beneficial for your better learning experience if you employ it effectively and for that you should start doing following things

i. Enact the character in notes wherever possible e.g., you are reading about Thomas Edison then you can think of yourself as Edison who is carrying out the series of experiments to reach to the final product i.e., the light bulb and you will definitely not be going to forget about him. You can also mimic your notes in different voices like Mickey Mouse or your favorite actor/actresses and you will be more involved in your learning process

ii. Exercise a bit that i.e., walk in your room while reading and this will increase your oxygen supply to the brain and you will feel energized and more alert to learn anything

iii. Try different positions while studying i.e., half an hour sitting in study room, then standing and then again sit or walk some quite place other than your study room. Make your comfortable schedule whichever works for you but remember pomodoro technique and follow it religiously

Sense of Smell: It will be little unusual for you but yes sense of smell will work for you in your learning process. Sense of smell is connected to the limbic system which controls our emotions, memory and creativity. The sense of smell is so prevailing that a just a slenderest redolence of smell can trigger your memory so deep that you can get information even from your early childhood. We call them as olfactory memories. Pleasant odor enhances our left frontal brain region which is responsible for our language related movement. Some scents are can work for you if you use it for atmosphere conception i.e., mocking a test environment in our study room and carrying that scent with us on the test day (if possible) it can help us recall lot of information. Some such scents are Rosemary (Fragrant evergreen herb) and Kouju (La Kouju- Fruity aroma made with Kouju cultivars) and Lavender. Rosemary scent increases our alertness, Kouju improves our memory and induce positive emotions and lavender soothes our nerves and increases concentration.

Sense of Taste: - When we taste something our taste buds stimulates activation of nerve receptors which send signals to our thalamus (controls consciousness) and cerebral cortex (controls mainly attention and memory). Whenever a matter of your favorite food arises you can't stop your mouth from watering even though you are not physically eating it but your strong sense of taste gives you that experience. We can have four major types of tastes sensed by our taste buds i.e., sweet, sour, bitter and salty, umami (glutamate or taste of broth). We can use these tastes for making our learning experiences more fun and better. You can have chewing gums like peppermint which can improve memory and reduce stress. Researchers have suggested that having a chewing gum at last moment of your exams can help you boosting your brain i.e., you can be focused in stressed situations. If it is allowed then keep a peppermint gum for your exams but it should be the familiar one i.e., you should have tasted that flavor at least ones

Meditation: -

How about that if I say that your IQ is not fixed and meditation can improve it to such a level that you can literally control your brain to what to think and when to think i.e., complete mastery over your thought process. Yes, it is undeniably possible just by meditating 25 minutes per day.

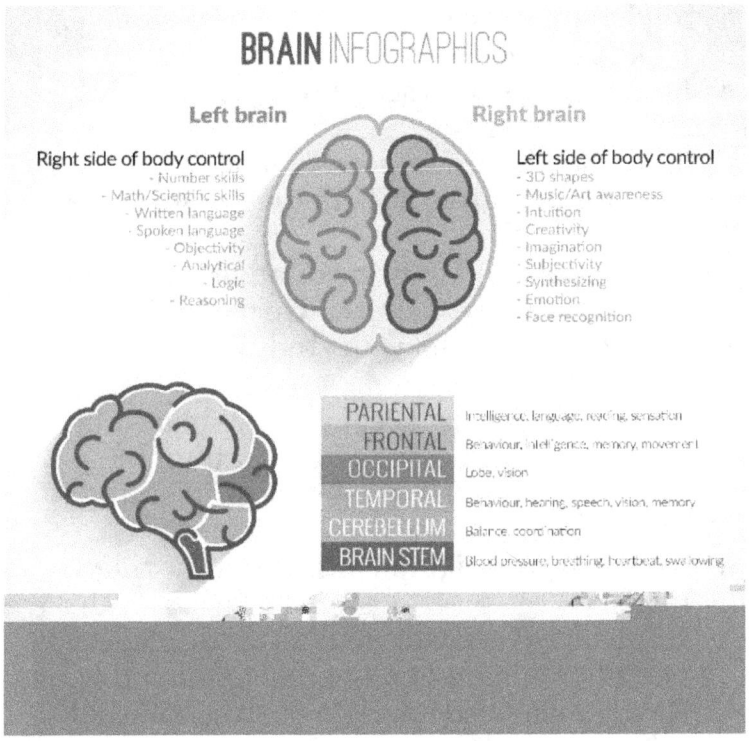

Meditation awakens the person within us which is calm, composed, smart, artistic, focused, insightful

and full of utmost attentiveness. Once you will understand, what meditation can do to your brain, you will definitely include this in your daily routine.

    i. Meditation makes our brain bigger by enabling whole brain synchronization. Benefits of this is it can boost our memory retention and intellectual thinking power
    ii. Meditation improves our memory recall process
    iii. Meditation amplifies our creativity
    iv. Mediation enhances Endorphins secretion i.e., feel good hormones in our brain
    v. Mediation increases Serotonin secretion i.e., make us happier
    vi. Mediation reduces secretion of Cortisol i.e., stress hormone
    vii. Mediation improves secretion of Melatonin i.e. hormone that gives us better restorative sleep

As we have understood what meditation can do for you, follow these steps of alternative nostril breath work before or in between your study session.

    i. Sit comfortably by keeping your spine straight
    ii. Seal your right nostril with your thumb
    iii. Inhale deeply with your left nostril for 4-5 seconds
    iv. Now seal your right nostril with your ring finger
    v. Hold for 2-3 seconds and exhale through right nostril
    vi. Next, inhale through right nostril, hold and exhale through left nostril
    vii. Do at least 15-20 rounds, by deepening the breath with each round

There are various other meditation techniques but it is recommended you should start with alternative nostril breath work and once you are comfortable go for Mindfulness, Zen, Kundalini yoga and Transcendental meditation.

We have understood which practices to stop and what are some new practices which we have to incorporate in our learning process. In next chapter we will understand various memory recall methods and their uses in our learning process.

Section II: Different Memory Training Techniques

Albert Einstein

We cannot solve our problems with the same thinking we used when we created them.

Different Memory Methods

In this section we will learn the different memory training techniques to commit our memory for recalling the information when we need it. Without proper armament it is difficult to win your war against forgetfulness. The methods mentioned here are those arms which will equip you to win in any situation where you need to recall lot of information it could be your exams, any presentation, daily items which you forget some or other while leaving for school or even if you are aspiring to become memory masters. I need one commitment for your side that you will start using these methods and I am confident that one of the methods will definitely work for you. If you have any misconceptions or myths about your memory just keep one thing in mind our brain has capacity of 2.5 petabytes which is sufficient to store videos which is enough for 300 years of play and you will still be left with some space. Start with one affirmation which I have recently learned and its working for me i.e. "I Am Enough". (20 Minutes 'I am Enough' Guided Meditation You Can Do Anywhere | Marisa Peer - YouTube).

Chapter III

Association and Pegging Method

This is the simplest method to start for even complete beginners who have just started to know, what is a memory training. In this method you have to associate whatever you already know which is a part of your Long-Term Memory, to any new information which you want to learn. We can understand this with the help of an example. Suppose you have to learn a list of 10 items and your list is as follows. If you will read books of various memory coaches earlier and wonder why everybody starts with a list, because our mind understands language of pictures than words i.e., our brain is an image processor so you have to feed it, in understandable binary code to get maximum benefits of it. Before starting any memory technique, I want to tell you that you have to imagine a lot of things and explore a minds creativity to its ultimate level. Sometimes you may feel these ideas are very childlike, but the fastest learners on earth are children. Also, whenever I you will hear words like imagine, visualize or suppose you have to run your creativity to such a level that you can see the items/objects/persons in front of. So close that you can touch them, feel them and can see as vivid as possible. Like I am telling you to suppose "Sun is in your house" then you should "Bright Orange Sun sitting on your couch, talking to you like a friend and giving you a positive energy". If your imagination power is at ultimate level these techniques will start helping you immediately. So, make our brains better and brighter, let us become child again and start learning new things.

So, here are with our very first list to start with,

1. Sun
2. Shoe
3. Color Wheel
4. Wisdom Teeth
5. Fingers
6. Dice
7. Continents
8. Spiders
9. Pregnancy
10. Metric System

By this list you could have got an idea what I am trying to tell you and if not, this is your clue, all the numbers are associated with the name, object or animal mentioned with them.

We have only one Sun.

We wear two pair of shoes

We have three basic colors in our color wheel i.e., Red, Blue and Yellow.

We all have four wisdom teeth

We all have five fingers in each hand and foot (With some exceptions)

We have seen six sides in dice cube and you certainly have played Ludo, Pictionary or any other games where you have used dice

We all have learned about 7 continents in our schools i.e., South America, North America, Europe, Africa, Asia, Australia (Oceania) and Antarctica

We all have seen spiders and we know they have eight pods i.e., they walk with eight feet

We definitely have known this that pregnancy period is 9 months (With some exceptions)

We all have studied metric system i.e., Meter, Centimeter, Kilometer and this is based on metric i.e., 10 times of the next measure like 1 Meter= 10 Centimeter, 1 Centimeter= 10 Millimeter etc.

We have made our basic associations and this will work for any list up to 10 items. We can put this in to test by taking another example i.e., list of 10 items

1. Amazon
2. Burn
3. Gate
4. Facebook
5. Space
6. Oil
7. Shares
8. Oracle
9. Ball
10. Zara

This is the list of world's top ten wealthiest people (data is of the dates, when this book was in writing). If you know the method it is as easy as pie. You will come to know within seconds and here is your association for world richest men.

1. **Sun- Amazon**

Visualize: Sun is rising and through sun's morning light you can see Amazon river which was due to darkness was not visible earlier as you see in movies. Amazon is formed by Jeff Bezos so Amazon is your clue for Jeff Bezos.

2. Shoe- Burn

You stop wearing leather shoes and only wear sports shoes as you get shoe burns i.e., burning sensation in your feet when you wear leather shoes. Burning sensation will remind you of Bernard Arnault, the billionaire #2

3. Color Wheel- Gate

Visualize: Password to open your college gate is three primary colors of color wheel. Gate will take you to Bill Gates

4. Wisdom Teeth- Facebook

Your father is telling you that you will get your Facebook account only after you get your wisdom teeth. Facebook is your clue for Mark Zuckerberg, Billionaire #4

5. Fingers- Space

Visualize: Your teacher is telling to your class that the name of space bodies should be on your fingertips. Space is your link to Elon Musk.

6. Dice – Oil

Visualize: Your dice of Pictionary game has fallen in oil tub and you hate touching that oil and hence cannot play further. Oil is you clue for Mukesh Ambani

7. Continents- Shares

Visualize: Your share broker is telling to your father the scheme of travelling all the continents and making money using different currencies. Shares are your clue for Warren Buffet.

8. Spiders- Oracle

Visualize: A spider is on your course book of Oracle and studying about web technology. Oracle will take you to Larry Ellison

9. Ball- Pregnancy

Visualize: You are studying about embryo formation in your biology and you read that embryo becomes like as rugby ball or pineapple till the 34th week of pregnancy. Ball is your clue to Steve Ballmer

10.   Metric System-Zara

Visualize: You are in a Zara Stores and asking for bill in yards where as they follow metric system only. Zara can take you to Armancio Ortega #10 Billionaire

The information associated with your list of 10 items which we call from now as Sun List is pegged i.e., we have used first information as hooks and the second information i.e., list of top ten billionaires as our garments which we peg on those hooks. Hope you found it easier than your previous method of rote memorization. You have to remember your first list that Sun list and any number of information having 10 items can be pegged and recalled using our Sun list. If you work a bit you can make your sun list as big as of

100 items and with little practice you can recall list of 100 items using your extended **Sun List**.

The third method which we are going to learn is **"Number Shape Method"** it is similar to Sun list but the list is of shapes resembling to our first 10 numbers. This is another list which you can make up to 100 numbers and you will have list of 200 items just and you can peg the information using that list. I will suggest to go slow i.e., first master the list of 10 items and then work of list of 20 items and then 30 items and this way i.e., slowly and steadily reach to the list of 100 items and 200 items as you feel comfortable. Here is your list of Number Shape method.

Numbers	Shapes
1(One)-Candle	
2(TWO)_Duck	
3(Three)-Heart	
4(Four)-Chair	
5(Five)-Hook	
6(Six)-Hockey Stick	
7(Seven)-Axe	
8(Eight)-Snowman	
9(Nine)-Balloon	
10(Ten)-Bat & Ball	

As you can see in the list the numbers are resembling to some shape and if you are a visual learner you can make use of this list effectively and make your associations in fun and easy way. We will learn one such example to understand our Number Shape Method in more detail. Suppose you have to memorize Biological classification of Living things i.e., Kingdom, Phylum, Class i.e., hierarchical classification can be done using number shape method. Our classification includes seven groups

1. Kingdom
2. Phylum
3. Class
4. Order
5. Family
6. Genus
7. Species

Once you associate with number shape method then recalling will become easier for you. Our association for this could be

1. Candle- Kingdom

Visualize: Candle is the symbol of your kingdom and it is must for everybody to keep a candle in home as there are lot of power tripping issue.

2. Duck-Phylum

Visualize: Duck as your accountant in school who manages all the files

3. Heart-Class

Visualize: Your class on topic heart and your instructor is so involved in the teaching that you can measure even your heart beats by his methods

4. Chair- Order

Visualize: Your grandfather is still the supreme authority and everybody obeys his order.

5. Hook- Family

Visualize: Your family has a unique hook which can even hang an elephant

6. Stick- Genus

Visualize: Your pair of jeans is an exclusive jean endorsed by world cup winning team of hockey players

7- Axe- Species

Visualize: You have seen a spaceship full of axes but alien has requested you not to tell anybody as they are not going to harm anyone.

We have numbers up to 10 to make associations, in later chapters we will learn numbers up to 20 or in some cases up to 30 for making our associations and pegging the information with them.

Our next technique is **Rhyming Number Technique** and this is also part of association technique only. As we have seen in the Number Shape method the shapes which are similar or look like numbers are used for association. Here we will use sounds similar or rhyming with our numbers and associate our list with them. This technique will work for everyone but for auditory learners this is one of the best. These is the fourth techniques of association method. So, by taking an example we can understand this technique. While going to school/college you may forget carrying something or other it may be your lunch box, cell phone or sometimes pencil box or even your ID card. Associate your items with rhyming number techniques and see if you can keep make a mental file of all your belongings.

Numbers	**Sounds**
1(One)	Bun
2(TWO)	Shoe
3(Three)	Tree
4(Four)	Door
5(Five)	Hive
6(Six)	Sticks
7(Seven)	Heaven
8(Eight)	Crate
9(Nine)	Line
10(Ten)	Den

Suppose, your compulsory list of items is:

1. Lunch Box
2. Cell Phone
3. Pencil Box
4. Id Card
5. Water Bottle
6. Glue Sticks
7. Planner
8. Stapler
9. Locker Keys
10. Wallet

We will associate this with the rhyming number list and make our compulsory list part of our long-term memory so that we will never forget these items again while leaving for school.

1. Association: Bun- Lunch Box

Visualization: Mummy has kept your favorite handmade bun in your lunch box which is unique ones and your friends started sharing their tiffin with you as they like sharing that bun between them.

2. Association: Shoe- Cell Phone

Visualization: Your teacher has instructed your class to keep cell phones to put on silent mode and keep in your shoes.

3. Association: Tree- Pencil Box

Visualization: Your pencil box is of the shape of a banyan tree and when you open it you can see erasers kept at branches and pencil and pens kept at trunk

4. Association: Door- ID Cards

Visualization: Your main door holds your hand and not letting you out without your Id Card

5. Association: Hive- Water Bottle

Visualization: You are giving water to drink to a worker honey bee and she is giving you honey in exchange

6. Association: Sticks- Glue Sticks

Visualization: Your hockey stick is broken and you are sticking it using your common glue

7. Association: Heaven- Planner

Visualization: In your dreams you are in heaven and angel is asking about your monthly schedule and you are checking your planner

8. Association: Crate- Stapler

Visualization: At your school fest, you responsible for distributing staplers to participants of a craft event

9. Association: Line- Locker Keys

Visualization: To open your locker you have kept a two-stage security, first is use locker keys and then type password "Draw a straight line"

10. Association: Den- Wallet

Visualization: You are in a zoo and your wallet has fallen in Lion's Den (Cage) and you are trying to convince lion to give it to you

Visualizations is trigger for our mind and awakens our creativity. Just by associating all your 10 items you will automatically get your necessary trigger of what all things you carrying and if something missing from your list or not. By this time, you will have idea how memory techniques work so, further we will dive deep into some concrete methods which will make your learning process exciting and fun. We will try using bigger lists and for that we will make our stories. So, our next methods are based on Story- telling or linking method.

Chapter IV

Chain Linking Method and Tagging Method

There are two types of story methods Chain Linking Method and Tagging Method. We will learn firstly about Chain **Linking Method.** In Chain Linking Method basic principle is you have to go on making connection between two objects and once you have connected the first object to the second one you should have to concentrate on second and third object only and once you connected second with third then your concentration should shift on third and fourth and thus chain goes on. Beauty of this method is your stories can be of any length which you can handle and also this method works in all situations that is you have to remember a list of items, a short paragraph or a long answer it will work for all of them. So, it is time to bring all your creative side in play and enjoy story-telling to your mind. We will understand the application of Chain- Linking Method by taking a paragraph from your Civics or Political Science book[iii].

Recently, politics have been classified into. **Democracies**, **Authoritarian** *and* **Totalitarian** *political systems. Democratic political systems are divided into* **three** *sub classes. The first of these has* **high** *sub system* **autonomy** *such as* **British** *or the* **American**. *The second sub class is characterized by* **limited** *sub system autonomy and it includes* **France** *of the Third and Fourth Republics,* **Italy** *after World War II and Weimar* **Germany**. *The third class of democratic political systems is made up of those with* **low** *sub system autonomy.* **Mexico** *provides an example of this type of democratic system. In **Authoritarian** systems liberty is restricted and **parliamentary** institutions are absent or meaningless, but they are not tyrannical regimes. The society is traditionally oriented and power is exercised by small groups such as, military leaders, bureaucrats, or religious figures. Examples are **Argentina** under Peron, **Spain** under Franco, and **Portugal** under Salazar. Finally, there are Totalitarian political systems as pre 1990 **Soviet Union**, the Fascist regime in Italy and the Nazi system in Germany. The regime in a Totalitarian system is based on a **dominant** leader supported by a mass party acting on an aggressive ideology which explains and conditions political actions. A Totalitarian regime is differentiated from an Authoritarian system by its total control over and attempt to regulate in detail on behavior and by the subordination of all organizations to the State*

We have to read this article and answer this question - what is the modern approach to classification of governments across the world? This is what happens in your classrooms but no, I am a different teacher and my approach is you should not be feeling overwhelmed after looking to such big paragraphs. Rather these paragraphs should be so interesting for you that you should dive deeper in the subject to have all the knowledge about that particular topic. So how does this happen?

Let us explore it with the different approach to your learning process. You have to note down all the highlighted key words from the paragraph and you have to tell your brain an altogether different but interesting story. Once you have told that story to your brain ask yourself if you can remember the whole concept and surprisingly you will remember. Our own interesting key words from the above paragraph are

1. Demo
2. Author
3. Tote Bag
4. Three
5. Height
6. Auto Ricksha
7. Big Ben
8. Statue of Liberty
9. Lime Soda
10. Eiffel Tower
11. Pizza
12. Tub Mummy
13. Law school
14. Mayan Riviera
15. Pearl
16. Lionel Messi
17. Picasso
18. Cristiano Ronaldo
19. Joseph Stalin
20. Dome

Once you are ready with your keywords you should build a story around it and by principle of chain linking method take two keywords at a time and make connections between them. Your story should be completely abstract, extravagant, interesting, outstanding, unusual so that you can see all the characters just by closing your eyes. Here is such a story for us.

You have to give a **Demo** of your world tour to an **Author** so that he can write a book for you. **Author** has **Tote bag** to carry all his writing materials. **Tote bag** has **Three** sections one for writing material, one for USB and camera and one for Laptop. Author's laptop is unusually big in **Height.** Its **Height** is as big as **Auto rickshaw.** This reminds you of hiring an **Auto rickshaw** when you have gone to visit **Big Ben** in London, England. At **Big Ben** you got a discount coupon to see to **Statue of Liberty** in half the price which you have estimated. You bought **Lime Soda** at **Statue of Liberty.** Along with **Lime Soda** you got another discount coupon for Europe tour and you decided to visit **Eiffel Tower** first. At **Eiffel Tower** you had **Pizza** at an Italian restaurant whose owner's **mummy** was selling **tub** in Paris and from there they go on to build the pizza food chain. Pizza Food Chain owner was actually planned to go to **Law School** as he was quite fascinated by a law school near **Mayan Riviera** in Mexico. While leaving his restaurant he has shown you a **Pearl** which he has got from **Lionel Messi** for his achievement on football field and also told you if you want to meet Messi today then you have a chance to meet him as he is town and you can meet him at **Picasso** Art Gallery. But when you have reached to Picasso Art Gallery you have got a chance to meet Cristiano Ronaldo also and you were thanking the restaurant's owner for this. Ronaldo has given you a rare photo of Joseph Stalin. In the photo Joseph Stalin is standing in front of Dome in Moscow which become a memory which you are going to cherish for your life.

Wasn't it an interesting story? Yes, it is. Now at this particular stage, you are confused how does this story will help you remember the keywords from our paragraph on Different Types of Governments. I will solve that mystery and I am quite sure you will not get bored of reading any large paragraph after this.

So. Our first Keyword Demo is Democracy,

Author is Authoritarian,

Tote Bag is Totalitarian,

Height is for High sub system,

Auto Rickshaw for Autonomy,

Big Ben for British,

Statue of Liberty for American,

Lime Soda for Limited sub system autonomy

Eiffel Tower for France

Pizza for Italy

Tub and Mummy for Weimar Germany (1933- I will explained this in detail in later section),

Law School for Low sub system autonomy,

Mayan Riviera for Mexico,

Pearl for Parliament's role in Authoritarian system,

Lionel Messi for Argentina,

Picasso for Spain,

Cristiano Ronaldo for Portugal,

Joseph Stalin for Soviet Union,

And Lastly

Dome for Dominant leader

Hope now it feels relatable. I can tell you the whole story and association of the paragraph's key words from this one story. If still you are not able to relate keywords with your paragraph, by little more practice it will start making sense for you. Also, by using chain linking method for all your large paragraphs it will become as easy as pie for you. The story you make can be about your happy moments or some instances which are still fresh in your mind. If you have very few such happy moments, once you will start making stories like this you will start getting more such happy moments. Story making skills will give you more and more key words and those keywords once fed in your mental filing system can be used for learning variety of topics. Also, in chain linking method, you just need a trigger or your starting thread and whole story will start unravelling for you.

We will learn one more interesting concept i.e., **Tagging method** and by this learning state and their capitals, top ten rivers, top ten mountains, top ten skyscrapers etc. will become effortless task for you. It is similar to your chain linking method but here we will not be making longer stories.

We will start with the countries and their capitals. Rule for putting tag or label is to make a meaningful but totally unrelated word from the data you have and then make a short association between the two information.

1. **Armenia-Yerevan**

   Tag for Country- Army men
   Tag for Capital- Your Van
   Visualization Sentence- Army man is asking for your van to go home and meet his family

2. **Belarus-Minsk**

   Tag for Country- Bell Rust
   Tag for Capital- Mink blanket
   Visualization Sentence- You are keeping your school Bell in mink blanket to avoid it from rusting

3. **Croatia-Zagreb**

   Tag for Country- Crow Asia
   Tag for Capital- Jug
   Visualization Sentence- You are watching a crow which is going to Asia and it has a jug in its beak

4. **Dominica-Roseau**

   Tag for Country- Dominic
   Tag for Capital- Rose
   Visualization Sentence- Your friend Dominic has brought lot of roses for you

5. **Estonia-Tallinn**

   Tag for Country- East India
   Tag for Capital- Tally

Visualization Sentence- You are learning Tally software from East India Company

This is how you can make your tags for any two information and easily memorize that information by short mental visualization sentences. We can take one more example of Top Ten Tallest Skyscrapers in the world to understand in even better.

Sr. No.	Skyscraper	Height in Feet	Tag for Skyscraper	Tag for Height	Visualization Sentence
1.	Burj Khalifa	2716	Buzzer	Neck Dish	You have a **Buzzer** on your **Neck** which reminds you of your favorite **Dish**
2.	Shanghai Tower	2073	Song	Nose Comb	You are singing a **Song** which has nasal sound so you are using your **Nose** and **Comb** is your mic
3.	Makkah Royal Clock Tower	1972	Making	Tub Can	You are **Making** a small Tub **of** ice-cream but it was so tasty that

						you made a full **Can** of ice-cream
4.	Ping An Finance Centre	1965	Ping Pong		Tub Shell	You have got a **Ping-Pong** ball made of **Shell** and you have kept it in your **Tub**
5.	Lotte World Tower	1821	Latte		TV Net	You are watching a YouTube video of how to make **Latte** at home on your **TV** but your **Net** is not working properly
6.	One World Trade Centre	1776	One		Duck Cash	You are dreaming of **One Duck** who has lot of **Cash**
7.	Guangzhou CTF Financial Centre	1739	Ginger		Duck Mop	Your dog **Ginger** has made a **Duck** to **Mop** your house
8.	Tianjin CTF Financial	1739	Jinn		Duck Mop	You have a **Jinn** whose face

	Center Center				is like **Duck** and he does **Mop**ping whenever he is sad
9.	CITIC Tower	1731	City	Duck Mat	You are in **City** to buy **Duck** printed **Mat**
10.	Taipei 101	1667	Tea	Dish Jog	You are having your **Tea** in **Dish** while **Jog**ging

With this we have come to end of all small looking techniques and it is time to learn big league methods i.e., Memory palace, Number Sound Technique and Mind Mapping.

Chapter V

Mnemonics

Knowingly or unknowingly, you may already aware of Mnemonics as we all have learned this is school which is called Rhyming Mnemonics which we call it as "The days of the Month" poem.

Thirty days hath September,
April, June and November;
All the rest have thirty-one,
Excepting February alone.
Which only has but twenty-eight days clear
And twenty-nine in each leap year.

There are many types of Mnemonics like Name, Music, Order and Spelling mnemonics. It has been explained with example of each type and the advantage of this method is that you can make your own mnemonics as it is the simplest memory method and helps in almost all circumstances.

Name Mnemonics

You would have learned your seven colors of rainbow using a technique i.e., either **ROY G. BIV** or **VIBGYOR.** This is an example of Name Mnemonics. In name mnemonics either we take first letter of each word of the list we have and make a meaningful word so that we can give another image to our brain to recall it whenever we required or we form acronym of a long sentence, definitions or meaning to have a small one word for recalling. Some more examples of name mnemonics are acronyms are

FANBOYS=**F**or, **A**nd, **N**or, **B**ut, **O**r **Y**et, **S**o (Coordinating Conjunctions)

HOMES=**H**uron, **O**ntario, **M**ichigan, **E**rie, **S**uperior (The Great Lakes)

PEDMAS=**P**arentheses, **E**xponents, **M**ultiply, **D**ivide, **A**dd, **S**ubtract (Order of Math operations)

LASER=**L**ight **A**mplification by **S**timulated **E**mission of **R**adiation

SCUBA= **S**elf-**C**ontained **U**nderwater **B**reathing **A**pparatus

Music Mnemonics
Music mnemonics is using rhythmic melodic templates which will have you engage more with your verbal memory tasks. Musical trainings have shown to change the brain structures so it is better to undergo at least one musical training i.e., learning any instrument as targeted training of verbal memory using music can have more instant effect on learning and verbal memory performance. Some examples of music mnemonics are

Every **G**ood **B**oy **D**oes **F**ine (Lines of Treble Staff from the bottom to the top)

Father **C**harles **G**oes **D**own **A**nd **E**nds **B**attle (Order of sharps and reverse it and you will get order of flats)
Great **B**ig **D**ogs **F**ight **A**nimals (Line notes of Bass clef from bottom to the top)

Happy **H**enry **L**ives **B**eside **B**oring **C**ottage, **N**ear **O**ur **F**riend **N**elly **N**ancy **M**ag. **A**llen **S**ally **P**atrick **S**tays

Close. **A**rthur **K**eeps **C**aring. (First 20 elements of Periodic Table)

Order Mnemonics

Order mnemonics is more helpful when we have to memorize the data in sequence. Some examples of this mnemonics are

Kids Prefer Cheese Over Fried Green Spinach (Kingdom, Phylum, Class, Order, Family, Genus, Species- Order of Taxonomy)

My very excited mother just served us noodles (Order of Planets- Mercury, Venus, Earth, Mars, Jupiter, Saturn, Uranus, Neptune)

May I have a large container of coffee? (Value of Pi- 3.1415927 May (3), I (1), have (4) ...and so on)

Spelling Mnemonics

Using spelling mnemonics, you can get rid of confusion of how a particular word is spelt. You can take all the alphabets and make your own sentence and you will never forget the spelling of that word again. Some examples of Spelling mnemonics are

ARITHMETIC: A rat in the house may eat the ice cream.

BECAUSE: Big elephants can always understand small elephants.

GEOGRAPHY: George's elderly old grandfather rode a pig home yesterday.

NECESSARY: Not every cat eats sardines. Some are really yummy.

OUGHT: Only unique guides have this.
RHYTHM: Rhythm helps your two hands move.

TOMORROW: Trails of my old red rose over window.

Chapter VI

Loci Method/Memory Palace Technique/Roman Room Method

As the name suggests this method is memorize long list of information i.e., up to 100 and 1000s of words. Using this method, you can memorize your points of speech, chapters and sometimes the complete book. Alex Mullen (USA), Johannes Mallow (Germany), Marwin Wallonius (Sweden), Jonas Vos Essen (Sweden), Simon Reinhard (Germany), Wang Feng (China), Sengesamadan Ulziikhutag (Mongolia), Zhen Aiqiang (China), Christian Shchafer (Germany), Yanjaa Wintersoul (Sweden), Ola Kare Risa (Norway), Boris Nikolan Konrad (Germany), Ben Pridmore (England), Prateek Yadav (India), Dominic O Brien (England) all these memory champions use this technique and YouTube is full with videos of how they do it. So, this method works for all and it should work for you too. We will put this method in work for you with some simple techniques and you will automatically start exploring the memory champion hidden inside you. One more astonishing fact these memory champions say that they have just an average memory and they have got results by training their memory with different methods.

To tell you the potential of this method the poet and famous lyricist Simonides of Ceos who was a prominent member of Greek society in 5th Century. He was in middle of a performance in banquet hall, when he got a message that two persons are waiting for him outside with a very important message and he came out of the banquet hall to meet them. At that time the roof of the banquet hall came crashing down and killed all the person present except these three i.e., Simonides and his two visitors. Now it was his responsibility to tell the people of town about their unrecognizable and dead loved ones and you can't believe he used a technique to memorize who was standing where while the performance was going on and he take each and every person to their dead relatives and to everyone's surprise he was exactly right. This was because of the technique called Loci (Lo-sigh) method. The story tells that there were no less than 300 people present at that time and can you believe is it possible for a human being to tell the exact place just by recalling. Yes, it was true for him and can be true for you also.

This technique says you can use the places which you go often to hide your items, nouns and whatever you want to remember and recall when it is actually required by you. The place could be your home, school, play ground or some historical place with lot of places which you know inside out. We know places in our home and if somebody asks, we can tell each and everything which present in our house and this is what helps us to associate many items in a list or keywords from points of speech for public speaking or long answers. This technique is suitable for associations more than 20 keywords or items and with little practice can be recalled effortlessly.

We will first see with a simplest example and then we will go to the little bigger list of 20-30 items so that you will feel comfortable working with it. Suppose you have to memorize types of triangles. There are six types of triangles which you have studied in your schools, three based on their angles are Acute triangle,

Right and Obtuse triangles and three based on their side lengths and they are Equilateral, Isosceles and Scalene triangles. We can use our three locations from our living room and three locations from our kitchen and we can keep record of these triangles in our long-term memory. In my living room those three places are couch, center table and tv and places in my kitchen are fridge, stove-top and microwave. We should take the locations which are in sequence so that we can mentally go to the places by saying at first place, at second place and so on. Now we will make our mental movie to learn the types of triangles. Our mental movie is as you enter your living room you found an **acute** safety pin on your couch and you take it and keep it on **right** side of center table, at the same time there was on your tv there was a program going on how to make **obtuse** angle triangles. Then you went to your kitchen and took out a slice of pizza which was in almost in **equilateral** (three equal sides) shape and you first kept it in pan to heat it on your stove top but when you started your stove top your pizza strangely started converted in to **isosceles** (two equal sides) triangled shape and you then put in to your microwave where it got converted in to **scalene** (all three sides of different lengths) shape and you ate it before any further change of shape happens. This is funny story of your coming home from school and having pizza and this will remind you of the six types of triangles.

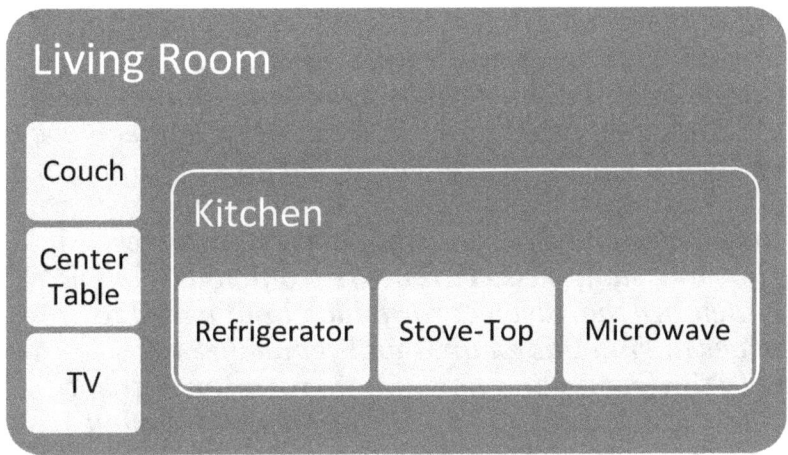

We will see this technique with one more example, which is points of speech, which you have prepared from your public speaking notes and your topic is "Your exam does not determine your worth". [iv]

***Final exams** do not determine my **worth**. No matter the grade – whether plus or minus, whether A or B or C or D or even F – it does not **connect** to the core of who I am in any way, shape, or form.*
*The **attitude** I have toward my final exams will directly affect how I **perform** on them. Therefore, I will do my very best, but when I sit down to take said exams, I will **breathe**, release, and let go of the outcome. I will realize that one test does not determine my entire **future**.*
*I will go with my **gut**. If I have listened, applied myself, and worked this **semester**, there is a good chance I will inherently know the answer to the question. If I do not, I will use my best skills and **strategies** to make a satisfactory educated guess. And it will all be okay if I get that answer wrong.*

*Education is a **gift**. Knowing not everyone in the world has such a **privilege**, I will not complain, whine, or spread negativity about my exams – even if I think they are unfair. There are thousands – even **millions** – of children who would give anything to take my unfair exam. I will take that into consideration before, during, and after my test. This, too, shall pass. **Ten** years from now I will not remember the answers for which I put pencil to paper during this exam. I will remember the **relationships** and meaningful **moments** spent over the expanse of the year because that's what I will choose to embrace. I will not allow one (or several or many) test(s) to ruin my day, my week, or my **summer**.*

*If, by some token, I made some drastic **mistakes** and choices that put me in a difficult position for this exam, I will choose to take **responsibility** for my actions and vow to make it better next time. I will realize that life offers me a plate every day, and that tomorrow is fresh with no mistakes. I will seize the opportunity to make **lemonade** out of lemons and do better next time.*

*Final exams do not determine my worth. I am **enough**. I have genius. Like Einstein said, I will not think I failed because I was a **fish** and they tested me on climbing a **tree**. I am not stupid. I may not be perfect in every area, but I have **value**. I have passions and **dreams** and they matter.*

*It is not okay to call myself a **failure**. If I fail, I will get back up and realize how I can do it differently next time but it does not make me a failure.*

*I am **unique**.*

*I have different **learning styles** and gifts than my fellow students.*

Not one of us is the same.

But we are all the same. No better than the other. I am not worth more than my **classmates** *and they are not worth more than me. I will take this exam with confidence knowing that whatever the* **outcome***, my worth will never change.*

We have taken out 30 key words and just by associating these keywords with our locations of school or home or other mental location and by simply mentally walking through those location we can recall them. We will do it by combining the locations of our home, way to school and supermarket.
My locations are: -

1. Main Gate
2. Parking
3. Wall Painting
4. Couch (Sofa)
5. Window
6. TV
7. Dining table
8. Bookshelf
9. Bedroom Door
10. Bed
11. Study Table
12. Closet
13. Family Photo
14. Grocery storage
15. Fridge
16. Wash Basin
17. Dish Washer
18. Stove Top
19. Microwave
20. Washing Machine

21. Bus-Stop
22. School Entrance
23. Security Cabin
24. Class-room
25. Desk
26. Black Board
27. Staff room
28. Accounts Office
29. Play Ground
30. Principal's Office

We will associate our keywords with the locations by the help of mental pictures or a chin linking story to recall it at the time of presentation. Here our story will be

You are coming home from a leisure trip and you saw a notice about your **final exams** on your **main gate**. While **parking** the car your dad was telling you it was fantastic trip **worth** every penny. When you entered your house, you saw **wall painting** has a **connection** with the place where you have gone for trip. You sat on your **couch** which was in no **attitude** as nobody was to sit on it from past many days at comfortably let you sit. Then you saw a dance **performance** through your **window**. You put on your **TV** and there was a man explaining how **breathing** exercises make our immune system stronger. Then your mother brings the food on **dining table** and your parents asking you about your **future** i.e., what you want to become. Somewhere inside your **gut** was saying that you are about to get a gift from your parents and you were right as you got your favorite book **"Unleash your Memory"** and now it has become part of your bookshelf (little bit self-promotion is ok, right). Then you open your

bedroom door to check your **semester** exam time table which you had pasted on it. Then you went to **bed** and started making your **strategies** about which subject to study first. Then you went to your **study table** to open your **gift** to check if it can help you with your exam preparation. Then you take out your medal from your closet which is source of your positive energy as it is a **privilege** for you, because you have got it from the hands of your favorite actor for extraordinary performance in exams. Then you turned towards your **family photo** which is worth **million** dollars for you. Then you go on to check your **grocery storage** if your mother has brought your **ten** important food items for you or not. Then you opened your **fridge** and found out all ten items there, this reminds you of the great **relationship** you have with your mother. Then you found empty tub of **moments** ice-cream in your wash basin. Then you started thinking about helping your mother in **dish washing** in your **summer** breaks. But you realized that it could be a **mistake** as you have burnt your fingers while using **stove top**. Then you have seen a pizza kept in microwave and you felt it is your **responsibility** to turn off the **microwave** and take pizza out. There was a lemon mistakenly lying on **washing machine** and you took that lemon to make **lemonade** juice for you. Then you got ready to go to school and reached to **bus-stop** but you found that it is still **enough** time left for the bus to come. When you reached to school's main entrance there was a big **fish** tank which was newly installed on main gate. Then your security officer came out of his **security cabin** and gives you a notice of **tree** planting project which has organized by local community for environment awareness week. In your **class-room** there were lot of new banners exhibiting the **value** of environment in our lives. Then you went to your **desk** where you got your project which was on Sleep patterns and types of **dreams**. Then you saw a quote on **black-board** which says "**Failure** is stepping stone to success". Then your professor called you in the **staff-room** to give you your **unique** identification number for inter school seminar. Then you went to your **accounts office** to buy questionnaire to find out your **learning style.** From there you went to **playground** to join your **classmates** who were waiting there for physical education class. In the meanwhile, you have got a message from **principal's office** regarding your **outcome** of national level paper presentation.

Hope this story will help you in preparing your speech for public speaking event and you will be able to tame that fear down. You should not worry if by any chance you have not understood it first time as it takes little practice to learn this method but once you got through with it you will become invincible with long answers or long list or any kind of presentations and you will start enjoying the process. There are many applications of this method which we will see in actual learning scenario in later chapters.

Chapter VII

Number Sound Technique/ Phonetic Number Sound Method

This is method is most useful when we have to memorize long list of numbers. Our brain does not understand the words or numbers it needs images for each information so the experts have given each number a sound and using that sound the images of numbers can be formed and used for our memorization. Our sounds for each number are as in the table below.

Number	Sounds
0	S(sa), Z(za)
1	D(da), T(ta)
2	N(na)
3	M(ma)
4	R(ra)
5	L(la)
6	CH (cha), J(ja), SH (sha)
7	K(ka), G(ga)
8	F(fa), V(va)
9	P(p), B(ba)

This method has few rules to use this sounds and those rules are
1. Consonants other than allotted to each number (as per above table) has no values while using the phonetic sound technique e.g., H, Q, S, W, X, Y
2. Vowels carries no values i.e. A, E, I, O, U are just to help in visualizing images
3. Repeated alphabets will be counted as one

4. Technique is completely based on phonetic that is what we hear is most important and alphabets values changes as per sounds e.g., in **Judge** though D (da) and G(ga) are there but it sounds like Ja so it will give us number 66(6-Ja, 6-Ja)

Using these rules our list of 100 numbers with all combination can be as per list below.

Before that we can try to understand how this technique works, suppose we have to memorize history dates i.e., when first world war started and ends then we have our dates as

Started on 28th July 1914- 06/28/1914(mmddyyyy)
Ends on- 11th Nov 1918- 11/11/1918(mmddyyyy)

So, we have our number sounds (using each number once only) for given data as

0-S,
6- Cha, Ja, Sha
2-Na
8-Fa, Va
1- Da, Ta,
9-Pa, Ba
4-Ra

If we will make a meaningful image out of these sounds then by combining two digits, we can visualize our images to make a meaningful sentence or a story.

06- Sa, Ja (Sage- Remember I have told you to concentrate on sounds only)

28- Na, Fa (Knife)

19- Ta, Ba (Tub)

14- Da, Ra (Door)

11- Ta, Da (Toad)

18- Ta, Va (TV)

So, our visualization for these images could be

Sage has taken out a Knife to cut a Tub to make a Door (06/28/1914)

A toad is searching another toad whom it saw on TV while bathing in a Tub (11/11/1918)

Does these dates are making sense for you now? Can you imagine the story vivid and clear? You have mastered the brain processing language i.e., image by visualizing numbers as images and now you can convert any number in to image but you need images for your numbers. We will see images for our first 100 numbers and this will help you in work out number of any digits. In this table number I have taken numbers starting from zero (0) and also given images for 00, 01-09. This will help you in converting numbers in two digits and then into images. You have to memorize these images which can be done by association, story or Memory palace method and you will become expert while working with numbers. We will see if we can work with bigger scary digit numbers or not. Here in India, we have 10-digit cell numbers and suppose your we have to memorize this number 8857034169 then as per technique we have to first break this number in to small chunks of two-digit numbers and our number can be broken down as

<u>88</u> <u>57</u> <u>03</u> <u>41</u> <u>69</u>

As per table of 0- 100 number our images for this number is 88- Five, 57- Log, 03- Sumo, 41-Rat and 69- Sheep. So, we can make our story with the images as Five wooden logs was kept to for making Sumo wrestling ring but one of the logs was unusable as a Rat has made a hole in that so in place of fifth log one Sheep was made to stand for time being. In a nutshell, you need to explore your creativity to the point so that you can start seeing those images in 3-D, you can smell those images, you can touch those images and you can talk to those images. Once you brought that imagination with your mental images nobody can stop you from becoming a real memory superhero.

Nos	Images	Nos	Images
0	See	47	Rack
00	Sauce	48	Roof
01	Suit	49	Rope
02	Sun	50	Lace
03	Sumo	51	Light
04	Sir	52	Lion
05	Sale	53	Loom
06	Sage	54	Lure
07	Sky	55	Lily
08	Sofa	56	Leash
09	Soap	57	Log
1	Tie	58	Leaf
2	Noah	59	Lip
3	May	60	Cheese
4	Ear	61	Shed
5	Law	62	Chain

6	Shoe	63	Jam
7	Cow	64	Chair
8	Ivy	65	Shell
9	Bee	66	Judge
10	Toes	67	Jog
11	Toad	68	Chef
12	Tin	69	Sheep
13	Dime	70	Case
14	Door	71	Cat
15	Doll	72	Can
16	Dish	73	Comb
17	Duck	74	Car
18	TV	75	Coal
19	Tub	76	Cash
20	Nose	77	Cake
21	Net	78	Cave
22	Nun	79	Cap
23	Gnome	80	Vase
24	Noir	81	Vet
25	Nail	82	Fan
26	Notch	83	Foam
27	Neck	84	Fur
28	Knife	85	Veil
29	Knob	86	Fish
30	Moose	87	Fog
31	Mat	88	Five
32	Moon	89	Fab
33	Mummy	90	Base
34	Mare	91	Boat
35	Mole	92	Bun
36	Match	93	Balm
37	Mug	94	Pear
38	Movies	95	Bowl
39	Mop	96	Bush
40	Rice	97	Book

41	Rat	98	Puff
42	Rain	99	Pop
43	Ram	100	Daisies
44	Rower		
45	Roll		
46	Roach		

We will work some scary numbers and try to check our mental imagination capabilities. Here we are with our golden ration which can be extended to more than 10,000 digits but we will try to memorize 100 digits

Golden Ratio φ = 1.6180339887 4989484820 4586834365 6381177203 0917980576 2862135448 6227052604 6281890244 9707207204 1893911374

So, here is the biggest digit which you may have worked with or even given a try to memorize in your learning process. Let us see if we can do it. Our images after taking two digits at time for this 100-digit monster are

Sr. No.	2-digit Number Chunks	Image
1	61	Shed
2	80	Vase
3	33	Mummy
4	98	Puff
5	87	Fog
6	49	Rope
7	89	Fab
8	48	Roof
9	48	Roof
10	20	Nose
11	45	Roll

12	86	Fish
13	83	Foam
14	43	Ram
15	65	Shell
16	63	Jam
17	81	Vet
18	17	Duck
19	72	Can
20	03	Sumo
21	09	Soap
22	17	Duck
23	98	Puff
24	05	Sale
25	76	Cash
26	28	Knife
27	62	Chain
28	13	Dime
29	54	Lure
30	48	Roof
31	62	Chain
32	27	Neck
33	05	Sale
34	26	Notch
35	04	Sir
36	62	Chain
37	81	Vet
38	89	Fab
39	02	Sun
40	44	Rower
41	97	Book
42	07	Sky
43	20	Nose
44	72	Can
45	04	Sir
46	18	Tv

47	93	Balm
48	91	Boat
49	13	Dime
50	74	Car

Now your first and foremost question could be why one need to memorize golden ratio at all. So, if you want to make another Mona Lisa then you need to memorize this golden ratio. Golden ration is very handy number that helps you to create beautiful, perfectly balanced designs that are aesthetically satisfying on a deep cerebral level. If you memorize first three digits then also, I can help you but you are planning to be a great artist it can work wonders for you. The trick to memorize this big digit number is to again make break it in to even smaller digestible chunks i.e., make 5 stories around it and you will be able to remember the whole 100-digit number. I want you to try it and make use of all the techniques you have learned in this chapter and come up with your unique set of stories.

Chapter VIII

Mind Maps®

According to Sir Tony Buzan the creator of Mind Maps®, it is simplistic and holistic thinking tool which activates our whole brain thinking engaging both the logical left brain and creative right brain using colorful visual diagram used to capture information on paper.

To create a Mind Map® you need

- A large sheet of plain white paper
- Minimum three colored pencils or pens
- Creativity
- A topic which you want to elaborate

A good mind map has three vital components a central image that depicts the main topic, thick branches radiating out from the central image which represents the key ideas each represented by different color and finally a main branch sprout subsidiary branches in form of second and third level branches which relate to further associated themes. Coincidently it resembles like neuron, axons and dendrites in our brain. Interesting and conceptual isn't it. Once we will start exploring with the help of example and it will become more clearer to us.

We will see what are some of the simple practices we should incorporate in our daily lives to start exploring our mental capacities with the help of Mind Maps®. Those practices are

1. Healthier nutrition- (Avocados, Broccoli, Blueberries, Extra Virgin Olive Oil, Eggs, Green Leafy Vegetables, Salmon Fish, Turmeric, Walnuts, Dark Chocolates)
2. Getting rid of Automated Negative Thoughts
3. Regular Exercise
4. Brain Supplements (Phytonutrients, Vitamin-B12, Omega-3, Antioxidants)
5. Positive Peer Groups
6. Clean Environment
7. Proper Sleep
8. Brain Protection from extreme sports and accidents
9. Learning New Skills
10. Stress Management

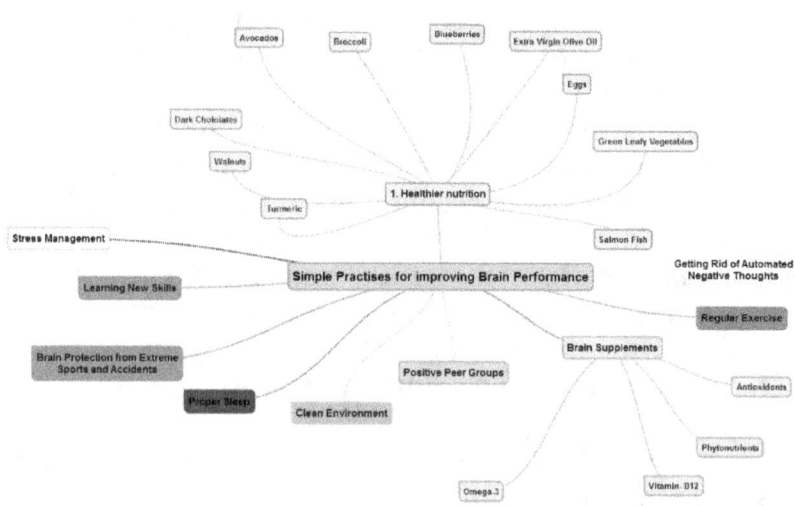

Similarly, the whole chapter can be summarized using mind map® and you can note down the noteworthy concepts and ideas and can refer whenever required. This is a handy tool for all your note-making both in class-rooms as well as at the time self-study. Till now we have learned the different concepts and now it is time to put our concepts in to action by involving real time scenarios from our syllabus and course books so that we can make utilize the memory techniques efficiently into our learning process. We will take examples of mostly all the subjects and topics which we study in our school and colleges so that memory training can become part of our daily routines. We will start with the most intimidating subject i.e., Mathematics and will see several topics from Physics, Chemistry, Biology, new languages, History, Geography, Arts, Music, and last topic will be some concepts from Computer Science.

Section III: A New Approach to Old Learning

Isaac Newton

What we know is a drop, what we don't know is an ocean.

Chapter IX

Mathematics

The Sumerians in southern Mesopotamia, which is now modern-day Iraq, developed a written language in about 3000 BCE. This was around the same time that they developed the first school of mathematics. People understand geometry and algebra by about 2000 BCE, by which time Sumer had become part of Babylon. Mathematics is the subject which is required in Physics, Chemistry, Biology, Astronomy, Computers, Robotics, Animation, Business and Commerce. If you are good at mathematics there are more chances that you can understand other subjects quite easily. If you have to memorize the great contributors like Hypatia, Thales, Pythagoras, Aryabhata, Brahmagupta, Bhaskara-I, Muhammad Al- Khwarizmi, Archimedes, Plato, Euclid, Diophantus, Khayyam, Bhaskara-II, Fibonacci, Yang-Hui, Copernicus, Galileo, Pascal, Euler, Lagrange and Ramanujan along with their timelines and inventions you can use combination of techniques like Loci method , Number sound method along with PNN Technique and you will come to know how much effort has been taken to make our life easier using mathematical principles and how everything is connected. Right from maths to science to history everything has some or other connection and your approach to study every subject will change and you will actually admiring the contributions of our ancestors i.e., without having the enough facilities we have today, they succeeded in achieving what they had put opted for without getting frustrated. The main branches of pure mathematics are: Arithmetic, Algebra, Geometry, Calculus, Trigonometry, Statistics and Probability.

As we have done in earlier sections will be start this chapter also with very basic information i.e., understanding different numbers and understanding those number types using our memory techniques. Broad Classification on hierarchical basis the different types of number are

1. Natural Numbers (All counting numbers except zero)
2. Whole Numbers (All counting numbers including zero)
3. Negative Numbers (Numbers less than zero)
4. Integers (Comprise of all negative numbers and whole numbers)
5. Fractions (any number which can be written in "a by b" format where b is not zero)
6. Rational Numbers (all integers and fractions)
7. Irrational Numbers (any non-perfect square root i.e., $\sqrt{2}$ or value of π)
8. Real Numbers (Both rational and irrational Numbers)
9. Imaginary Numbers (square root of negative numbers e.g., $\sqrt{-1}$ which can be written as *i* i.e., $\sqrt{-9}$ is $3i$ where $i = \sqrt{-1}$)
10. Complex Numbers (Both Real numbers and Imaginary Numbers)

This list can be memorized using association method. You can use Sun list, Number Shape Method or Rhyming Number method. We will try to learn using **Number Shape method** and don't worry we will learn in conceptually so that you can start playing with numbers and solving math will be one the fun activities for you.

So, once again we will make our story and our story is: You are in a supermarket to buy candles for your birthday and you asked store keeper for set of 10 candles and he said they won't keep any set which contains zero **(Natural Numbers)** in it. Then you went to another supermarket where they had all the combination which one can count and you got set of 10 candles but they were in Duck shape **(Whole Numbers).** You bought it and came out to visit your doctor for your heart check-up and doctor says there is negative **(Negative Numbers)** chances of getting any disease as you are taking good care of yourself. You took your negative report and candles and kept in your bag at one place and sit on chair in his waiting room **(Integers)** as doctor has told you that you should divide your daily activities and you can keep all your worries of the hook **(Fractions).** He has also recommended certain healthy foods. You took his recommended diet sheet chart and kept that to in the bag along with candles and negative report and started for your friend's house to collect your hockey stick **(Rational Numbers)** for weekend hockey match in your locality. Your friend called you and asked you to come to pie-point **(π- Irrational Number)** with the axe as he has to chop extra portion of a tree which is blocking playing area. Then you took packed hockey stick and kept in the same large bag which is actually in a shape of a snowman

(**Real Numbers**) and you started moving towards your home. On your way you saw a big balloon in the sky with motivational quote "everything is possible", seeing which you have started imagining (**Imaginary Numbers**) that you have won your hockey match. But when you reached home and checked and found your friend has given you a bat and a ball in place of hockey stick and you are kind of a complex (**Complex Number**) situation.

This story will remind you of the 10 number types. I will suggest you to first work on basic concepts and then once your foundation is strong, you can start handling all the math problems at ease.

We will now start taking out all our fears step wise by learning concepts of each mathematics section and why solving math problems could be fun. We will first understand about arithmetic. Arithmetic is Number art ("Arith" means Number and "Metic" is Art in Greek) and it will test your artistic skills i.e., how creative you can be with the numbers and arithmetic will become a memory sport for you. The basis of arithmetic is our four basic operations i.e., addition, subtraction, multiplication and division and we all know the notations of these four pillars and but do you the know the fact by knowing only two i.e., addition and multiplication you can recognize other two. How let us understand with help of an example.

If you say 6+10= 16, it is same as 10+6=16 and in subtraction it will become 16-6= 10 so you can say Addition and subtraction are inverse operations. Same is the case for multiplication and division i.e., multiplying 9 with 18 we will get 162 or 18 with 9 will also get 162, and when we reverse it 162 divides by 18, we will get 9 or 162 divides by 9 will give us 18. So, if you associate these four operations you can start thinking differently about them. Also, we know mnemonics to solve the order of operations and our mnemonics is

PEDMAS=Parentheses, Exponents and Square Roots, Division, Multiplication, Addition, Subtraction.

Or

BODMAS= Bracket, Order (Power), Division, Multiplication, Addition, Subtraction.

Whichever you find easier you should use that method. Now we have understood basic concepts so let us dive deeper in to complex arithmetic problems and see if our newly formed concepts are helping us or not. We will make use of distributive property to break our make our complex problems in to small manageable chunks.

Suppose you have to solve 19x729 which can also be written as 19x (700+29).

If you want to simplify even further it can be (20-1) x (700+20+9). Can you solve this now? Does it look less intimidating. No, let us see

20X700= 14000,

20X20=400

20X9=180

Then add these three products and you will get 14580

Then, 1x700= 700

1X20=20

1X9=9

Which is our original number 729, now subtract 729 from 14580 and your final answer is 13581.

This method can be used in finding out our squares and square roots also. Let us take out squares number 67 using distribution and association technique

67x67 can be broken down in to (60+7) x (60+7) and we can find our three components as 3600+840+49 and after solving this we will get 4489 which is our square of number 67. Likewise, for 89 we can solve as (90-1) x (90-1) and we will get our components as (8100-180+1) and our final answer will be 8921 which is square of 89.

Hope you have already start loving your calculations as your perception of looking towards math problems are started changing. This new approach of looking towards complex problems will solve your all the learning difficulties as these are myths and if you can believe it as easy tasks it will automatically stop intimidating you. Let us check our method for problems of square roots also. Suppose you have to take out square root of 5329 and to find out this we should know first our squares of 0-9 numbers which we can call alphabets of number system (as these are the basic components on which the whole number system exists) and our squares of 0-9 numbers are

$0^2 = 0$

$1^2 = 1$

$2^2 = 4$

$3^2 = 9$

$4^2 = 16$

$5^2 = 25$

$6^2 = 36$

$7^2 = 49$

$8^2 = 64$

$9^2 = 81$

Now we have to concentrate on unit numbers of these squares of 0-9 numbers for unit digit of our square root numbers and let us see some rules to calculate the square root

1. Break the square number in equal parts (in case you 576 as your starting number then you should break it as 05 and 76)
2. Check the unit digit of right-hand side (RHS)'s number (e.g., if your unit digit of square number is 4 our unit digit will be either 2 or 8 from our squares of 0-9 numbers)
3. Now concentrate on left hand side (LHS) number and check it is falling between squares of which two numbers between 0-9 and take the lower number as your first number
4. Finally multiply the two probable numbers we have got for our left side number and check if the product is less than or greater than our left side number. If it is lesser than the left side number you should use lower number as unit digit and if it greater then higher number as

unit place. I will explain it by taking out the square root of Number 5329.

Step 1- Breaking the number in two equal parts 53 and 29.

Step 2- Unit digit of number at right side is 9 so our unit digit will be either 3 or 7.

Step 3- Our number at LHS is 53 which falls between squares of 7 and 8 i.e., 49 and 64 and we have to take the lower number so our tens digit in this case is 7.

Step 4- Now we have to multiply our probable numbers which we have got for our LHS which is 7 and 8 and their product is 56 which is greater than 53 so our unit digit will also be the lower number between 3 and 7. So, our unit digit in this case is 3.

Step 5- So our final answer is square root of 5329 is 73.

Once you will practice these steps for other numbers it will become matter of seconds to find out square root of larger number also. To make it even simpler try to take out squares and square roots of numbers up to 0-30 and you can start managing square of up to 0-999 and square roots up to 6- digits numbers. So, I start taking math as a game and make your own rules to play that game and broaden your horizon till the level you want.

As we have learnt some basic concepts from Number Art i.e., Arithmetic, now will try to acquire some basic skills for another nerve-wracking subject i.e., Algebra and we have to believe the best definition on web "Algebra is the study of mathematical symbols and rules for the manipulating these symbols". Please concentrate on the word manipulate that means you have total control over this subject if you give a little bit of focus on the concepts. Let us understand with one example which will initially seems little difficult but once you will understand essence you will come to know how easy it is. Our problem is

Problem: I have two fields that total 3200 hectares (1 hectare = 0.01 Square Kms). Yields for each are 2/3 quintals (1 quintal =100 kgs) of grains per hectare and 1/4 quintals of grains per hectare. The first field gave 700 more quintals than the second. What are the areas of each field?

Solution: How you will start solving this problem is let area of first field is "x" and area of second field is "y" and

$x + y = 3200$

$2/3x - 1/4y = 700$

After solving it further and using lot of calculations on paper we will reach to the solution as.

$x = 1800$ and $y = 600$

But the better approach would be,

$2/3x - 1/4(3200-x) = 700$

$2/3x - 800 - 1/4x = 700$

$(6-3)/4x = 800 + 700$

$3/4x = 1500$

$3x = 6000$

$x = 1800$ hectares

so, $y = 2400 - 1800 = 600$ hectares

and beauty of this approach is you can solve this in your mind without using pen and paper which was scarcity in past and people were doing these calculations mostly in their mind. Till this time, I think you will be convinced that everything we learn is not to torment us but to make us stronger to deal with the problems and finish it then and there it self but our approach is what actually limit us in achieving it. If you are not convinced, I have many more ideas to justify this. We are at our second topic only and when you will reach to the end of this book, your concept of learning process either would have changed or you will be in research mode of your favorite topic as to "How can I do it like a pro?"

With this let us see one more interesting example to validate our point. Before that some history do you know from where the word algebra came it was from the book "Hisab **al-Jabr**-W'al Muqabla" written by 9th Century (Around 820 CE) Persian Mathematician Muhammad Ibn Musa Al- Khwarizmi. This is the finding by most of the scientist and mathematicians and its English version is The Compendious Book on Completion and Balancing and if we see dictionary word for Compendious it is Compact and the word "Al-Jabr" means restoration so it is to reduce or repair your problem and not to increase it. So, in short algebra is related to mostly completion and balancing and restoration of problems to least possible level and applications related to this. If you want to learn the complete history that is from 780 BCE Babylonian times, to concept of finding value of x in 598 BCE by Indian Mathematician Bhaskara and finding fractions by Samudragupta, to taking our surface and volume of Sphere by Archimedes in 287 BCE, (+) and (−) signs by French Mathematician Francois Viète in 1540, to biggest milestones i.e. getting its actual name and bringing this subject for the help of humanity has ties by Diophantus and Muhammad Ibn Musa Al-Khwarizmi all can be learned and memorized using Number Sound Technique so that the concepts start getting clear in our mind as to how and what have been introduced to this field and what was the actual purpose of that introduction . Learning Algebra helps us to develop our critical thinking skills such as problem solving, logical reasoning and learning patterns. Whatever equation we solve be it from our science or math they all have their roots in algebra. So, algebra can help you in many ways if you start enjoying it.

Problem: What is unique about number 8549176320?

Solution:

Step 1: Think, what could be unique about this number?

Step 2: It starts with Eight then Second Number is Five and third Number is Four and then Fifth Number is Nine

Step 3: *Is there a pattern?*

Step 4: *Yes, it has a pattern, it is arranged in Alphabetical order (**E**ight, **Fi**ve, **Fo**ur, **N**ine, **O**ne, **Se**ven, **Si**x, **Th**ree, **Tw**o, **Z**ero)*

Step 5: *Is something more unique about this?*

Step 6: *Should I try dividing with each number?*

Step 7: *Bingo, it can be divided by all the units i.e., 0-9 numbers expects 7.*

Got the point, it's all about approach and you can solve any problem, isn't it?

It is time to tame the devil of **Geometry**. Geometry word has Geo means Earth and Metry from metron means measurement. It short if we have to measure anything which is on earth i.e., all the shapes which mainly are point, line and plane it will come under Geometry. Also do you know geometry exists from almost 2500 BCE and Egyptian pyramids are made using master planning of geometrical principles? Then comes Herons Formula of mapping the area of triangle (2000 BC) followed by Pythagorean Triples (600 BC) and then came Archimedes with Area and Circle and Volume and Surface area of sphere and (287BC). This all led to the Modern era of Geometry starting from 1596 and Co-ordinate planes came in to existence and today we have reached to fractal geometry and today's buzz word is 248-dimensional symmetry i.e., E8 lattice which was introduced to us in year 2010. Once I will tell you some interesting association then you will start liking this subject. Our first such association is Point, just connect two end points and you will get a line, Connect two different points to one common point you get angle, Connect all three points with each other you get triangle, connect four such point and your quadrangle is ready, then you have pentagon, hexagon and you can go up to megagon(*mega meaning great*) or 1 000 000-gon which actually defined till date and then we have stopped counting and we have started saying all higher shapes which we mostly don't use as polygon. Coming to our shape line, we have arrows, midpoints, endpoint, then based on angles we have right angle, obtuse angle and acute angle. We also have straight angle (180°) and Reflex angle (>180°) and followed by planes and the list goes on. Geometric figures are completely associated with each other and are easy to learn if you follow association technique. We will solve one problem and try to understand as to whatever I have told you are actually have any proof or not. We will take our area of a square embedded between four circles.

Problem: Four circles of equal radii are in a square box and another square is embedded between all the four circles. The side length of bigger square is 4 inches so what will be the area of the inner square.

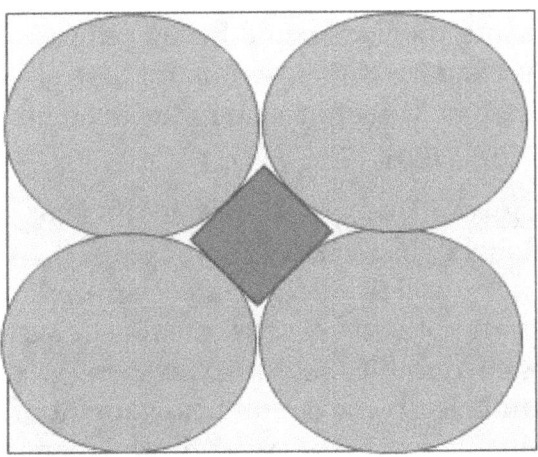

Solution:

Can answer be told just by looking at the figure? As one can tell the solution in just three steps.

Step 1: Diameter of each circle is 2 inches and surface area of each circle is same as their radii are equal. So, Radius of each circle is 1 inch.

Step 2: Follow Pythagoras theorem and you will get Side of inner square.

So, from Pythagoras theorem hypotenuse of outer square is $a^2+b^2=c^2$

Here C is our hypotenuse So $c=\sqrt{a^2+b^2}=\sqrt{32}= 4\sqrt{2}$

Diameter of circle is 2 inches

Side of inner square= $4\sqrt{2}-4=\sqrt{2}-1$

So are of inner circle= $(\sqrt{2}-1)^2=2-2\sqrt{2}+1=3-2\sqrt{2}=0.1715$ Square Inches

If we would follow our regular procedure then we would have first taken out area of square and then area of four circles and then we would have tried to find out are of inner square isn't it. Again, math is all about approach i.e., how you approach to the problem and it will become your buddy.

Coming to **Calculus** our first question could be why do we study calculus. Logical answer is studying calculus will help our mind to understand scientific method of analysis and practical use is calculus helps us understand the astronomical bodies, weather patterns, electric and electronic circuits, movement of sound and light and what not. This will tell you about the opportunities we have in this word if we devote complete concentration on a subject, we like the most. Just by exploring few pages on google search or YouTube or wherever you search for your answers on web you will find every field has ample opportunities it is we who have to explore and become what we want to become. On this note let us understand the origin of calculus. Yes, it is the same old story we all have heard i.e., Newton and apple tree. But Newton has found gravity but was not knowing how to calculate why the speed of falling object keep on increasing as it goes nearer to earth which we also called instantaneous rate of change of speed in technical language. Sir Isaac Newton with the help of German Philosopher Gottfried Leibniz. There are still debates going on and there are records that that it has been founded by Indians in year 1350 i.e., almost 250 years before Newton and research on the same is still going on. We will come back how this can simplified so that our sole purpose of understanding it better and put in to application. How one can calculate area or speed of an object whose shape is continuously changing. In that case we need to understand pattern and we are sorted. We do it with the help of slope and coordinates and as the slope moves values keep changes. With the help of some formulas, it can be determined. We have to keep it mind if we are working with definite or indefinite slopes. To simplify this, I will tell you with example as to how a typical calculus problem are so easy to understand. Before that let us note down and memorize all the calculus type. If we have to define, there are only two types of calculus, one when dividing in small pieces and calculating change from one moment to next is **Differential calculus (Derivatives)** and technically it is called instantaneous rate of change and slopes of the curves. The other is type is about integrating (i.e., joining) things together like areas under or between curves and called **Integral calculus (Integration)** and technically called accumulation of quantities and areas under or between curves when we plot a graph. This will clear your basic concept of calculus. The whole concept is based on three basic pillars limits i.e., predicting value of a function at any given point, derivatives i.e., taking out rate of change of a function and integration i.e., calculating area under the slope or curve.

To summarize it fully the whole concept of calculus is just translating the odd shapes into language of math. For example, if you are bouncing a rubber ball and thinking that the speed at which it touches the ground is constant but actually it is not and if you to calculate this change you will have to note down distance travelled by the ball and time taken to travel that distance (Speed= Distance travelled/ Time Taken).

That instantaneous change in distance and time taken to cover that distance is plotted on graph and average speed is calculated.

Suppose you are doing catching practice with your friend, using the same ball. If you have to calculate what is the maximum distance that the ball can cover, for that you need the angle and speed at which the ball is thrown, at same angle and different speeds or at different angles and at same speed you have variety of distance covered by the ball. This calculation of different distances is what is done by calculus. The same principle is used for calculation of area under the trajectory of ball thrown between you and your friend. Hope by now you have the concept of calculus as to why we have to study calculus.

We will move on to new topic i.e., **Trigonometry**. The word has two parts Trigonon i.e., **triangle** and metron is **measure i.e.,** study of calculations related to the sides and angles of triangle. When Egyptian were building pyramids (dating back in 1800BCE), which can easily be represented as triangle from front, they need to measure the sides of the triangles to each side to form a balance pyramid. They did this by drawing a circle. They were knowing the radius of circle that has become one side of triangle for remaining two sides they had to measure the length so for that they coined a word sinus i.e., band or chord (which is modern day sine). Sine and cosine are actually discovered by Aryabhata (476 CE- 550 CE). Muhammad ibn Musa al-Khwarizmi (780 CE- 850 CE) come up with tangent and Abu Wafa' Buzjani (940 CE- 988 CE) brought secant, cotangent and s cosecant. Now come back to our story of Egyptians i.e., how they calculated the other two sides of triangle. They had used angle and length of opposite sides and hypotenuse by plotting right angled triangle. Mind it the pyramid is not in right angle triangle. They had plotted a right-angle triangle inside the base triangle to divide it in to two triangles. They were knowing value of one angle and by plotting right angle they have values of two angles. You already know they were knowing length of one side so they come up with a formula in which they used by using angle and distance as their base. As we all know for a given angle the ratios of side remain the same no matter how big or small a triangle is. So, by using all these data they come up with following formula

Ratio 1= Opposite side/Hypotenuse (Which is our $Sin\theta$)

Ratio 2= Adjacent Side/ Hypotenuse (Which is our $Cos\theta$)

So, for example they had their hypotenuse angled at $45°$ and length of the hypotenuse is 100 mtrs, then they had enough data to calculate the length of remaining two sides.

We will solve here for them to understand it better.

Sin 45°= Opposite side/100

$1/\sqrt{2}$= Opposite Side/100

Opposite Side= $100/\sqrt{2}$=70.71 Mtrs

Similarly

Cos 45° = Adjacent side/100

$1/\sqrt{2}$ = Adjacent side/100

Adjacent side = *100/√2=70.71 Mtrs*

So, we have helped Egyptians to find out the length of the other two sides of triangle in a pyramid. Actually, it is other way round i.e., they have helped us determine any unknown distance with minimum of data by the use of triangle no matter how astronomical the figure is. Now let us go in to some more basic information as we have seen $Sin\theta$ and $Cos\theta$. There has to be some association between these two and that association is

$Sin^2\theta + Cos^2\theta = 1$,

All other trigonometric formulas are based on this on relation. First Aryabhata gave us Sine (*jya-relationship between half an angle and half a chord*) and Cosine(*kojya*) in 4th and 5th Century AD. Remember Egyptians only used it and not given the name as sine and cosine. Actually, Aryabhata measured the sine and cosine tables for us in 3.75° intervals from 0° to 90°. Then Varahmihira brought the relation $Sin^2\theta + Cos^2\theta = 1$ using Aryabhata's tables. Now how to remember the basic relationship between sine, cosine, tangent, cotangent, secant and cosecant. Here also make an image for all the relationships and you will have ready trigger for your memory recall. That trigger is trigonometric hexagon and It look like as below.

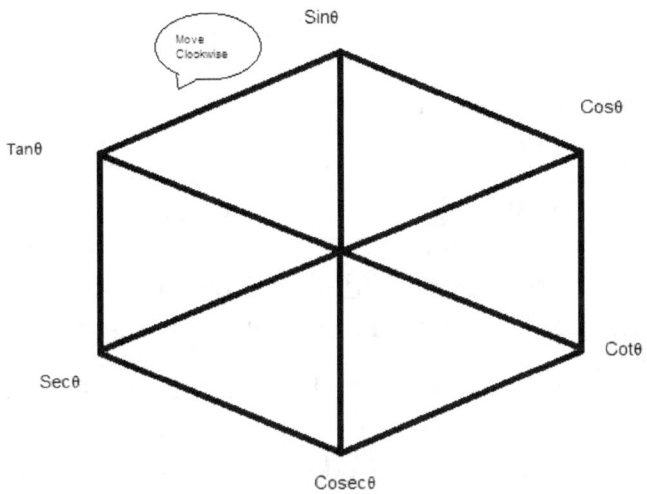

Move Clockwise:

1. Tan θ= Sin θ/Cos θ
2. Sin θ= Cos θ/Cot θ
3. Cos θ= Cot θ/ Cosec θ
4. Cot θ= Cosec θ/ Sec θ
5. Cosec θ= Sec θ/ Tan θ
6. Sec θ= Tan θ/Sin θ

Exactly opposite:

1. Tan θ= 1/Cot θ
2. Sin θ= 1/Cosec θ
3. Cos θ= 1/ Sec θ
4. Cot θ= 1/Tan θ
5. Cosec θ= 1/ Sin θ
6. Sec θ= 1/Cos θ

Look Front:

1. Tan θ= Cot (90- θ)
2. Sin θ= Cos (90- θ)
3. Sec θ= Cosec (90- θ)

Use your imagination power and you can come up with almost all the associations using this one simple hexagon method. This should make your life little bit easier while dealing with trigonometric problems.

Now we will see another topic that is **Statistics** which is derived from Latin word *Statisticum Collegium* meaning "Council of state" and Italian word *Statista* meaning "Statesmen". It is Statistics which is a branch of mathematics dealing with the collection, analysis, interpretation and presentation of masses of numerical data or a collection of quantitative data. To scientific right, now in general terms Statistics is all about collecting data and analyzing it. For example, how many students across the world likes Statistics is comes under the scope of Statistics. As statistics have lot of components various contributors are found in this field and it started from Ancient Romans used census for agriculture in 4500 BCE, Then in China population census was done in around 3000BCE, Roman emperor Augustus conducted surveys on birth and death of his citizens in 27 BCE. John Graunt a first known English statistician, found its use in demography in 1620. Then it is further found contributions from William Petty, Thomas Bayes, Gottfried Achenwall, Blaise Pascal, Pierre De Fermat, Daniel Bernoulli, Pierre-Simon Laplace, Carl Freidrich Gauss, Adolphe Quetelet, Francis Galton,

Karl Pearson, Charles Spearman, Wesley Mitchell, William Gosset(Penname-Student), Kristine Smith, Ronald Fischer, Harold Jeffreys, Franck Wilcoxon, P C Mahalanobis, Jerzy Neyman, R C Bose, Frank Yates, Abraham Wald, Andrey Kolmogorov, Maurice George Kendall, J. Wolfowitz, PV Sukhatme and list goes on. I will recommend to search them over internet and try to find out their contributions to the field of statistics. This will help you to inculcate the research methodology and you will find quantum of resources available freely for your academic use. Why am I saying this because once you start analyzing your data in your own research you start learning typical statistical patterns. What is its application? It is mainly used for definiteness (Confirmed Information), condensation (presenting data in lucid format), comparison, formulating and testing hypothesis (used for research purpose) and forecasting (e.g., how a product will sell based on past performance). When you will start exploring this topic you will find it present everywhere, as whenever someone need to forecast, control or explore any kind of data they need statistical expertise. Once you master this topic, you can make your career in any industry or sector you want, as it has its application in almost all the fields. Data scientist is in high demand and is going to remain hottest job of 21st Century.

Statistics is based on four basic components

- Formulating Questions
- Collecting Data
- Organizing and Summarizing the data
- Making Conclusions

To memorize these four steps your story could be – You are on an agriculture farm and questioning the farmer what is the procedure of sowing a seed. He tells you different possibilities of sowing different seeds season wise, area wise and irrigation schedule wise, you have noted down all the data and organized it in different columns based on types of crops, seasons, irrigation schedule and all possible organizations. Then you wrote your summary for your academic project and submitted your project after including conclusions. This is your story for memorizing the four steps and also your experience with data collection process in Statistics.

We will learn through one concept of statistics to know how calmly a statistics problem can be tackled and befriended. The five most important methods of statistical data analysis are Mean (Average), Standard deviation, Regression, Sample size determination and Hypothesis testing.

We will take a real-life example of standard deviation problem. Take the example of weather forecasting. You watch news and sometime you wonder how they predict the almost exact weather condition for next 2-3 days. We will try to learn how a news channel used standard deviation as a tool.

Problem*: A news channel has temperature data on 9th Nov 2020 of 10 cities of world and help them with the standard deviation of temperature using the below given data series. The data is as in table given below*

Sr. No.	Cities	Temperature (in degC)
1	New York	17°C
2	Beijing	6°C
3	New Delhi	12°C
4	Tokyo	15°C
5	Berlin	5°C
6	Moscow	1°C
7	Jakarta	26°C
8	Brasilia	25°C
9	London	12°C
10	Paris	11°C

Solution: We can take out standard deviation in 3 easy steps.

Step1: Calculate mean

$x = (17+6+12+15+5+1+26+25+12+11)/10$

$x = 140/10 = 14$

Step2: Calculate the difference from mean and taking its square

$\sigma^2 = [(17^2+6^2+12^2+15^2+5^2+1^2+26^2+25^2+12^2+11^2)/10] - 14^2$

$\sigma^2 = 228.6 - 196 = 32.6$

Step3: So, our standard deviation is (by taking out square root)

$\sigma = 5.71$

So, temperature is 5.71 point deviated from the mean i.e., 14. When the deviation is large that means scores are more widely spread out from the mean and vice versa.

This brings us to the last topic of maths i.e., **Probability** which is actually characterized by randomness and uncertainty. A gamblers dispute in 1654 led to the mathematical theory of probability by two famous French mathematicians, Blaise Pascal and Pierre De Fermat. The game consisted in throwing a pair of dice 24 times; the problem was to decide to either bet or not for at least one occurrent of "double six" during the 24 throws. Before that in 15th centuries also there is mention of few special problems solved on games of chance by Italian mathematicians (1560- Gerolamo Cordano). Its application like statistics has everywhere i.e., wherever you need to do any kind of modelling you need probability. IN general terms wherever you hear the word chance you are referring to mathematical term probability it is as simple as that. To simplify this for you please note only two rules any probability will be between 0 and 1 i.e., it can be zero and it is either be one or less than one in all situations. We all take insurance policies; you know insurance business is totally based on probability. It works on law of large numbers i.e., more the policy holders less is the chances of declining the policy claims as predictive power of insurance companies increases. This is the case for almost all fields but insurance companies can be considered as one of the case studies.

Let us understand with some real-life examples and you will understand the terms like 50% chance of 0.5 probability.

1. When you toss a coin there are only two chances i.e., either you will get head or tails so

it become 50% chance and probability is P=1/2=0.5 which is less than one
2. If you are playing any game of dice (die-singular) then there could at least one chance of getting number 6 if dice is thrown for 6 times so here your probability is 1/6 i.e., P=1/6 which is also less than one
3. Little trickier problem could be there are three different colors balls in a bag blue 5, red 6 and green 4 so what are our chances of getting red balls. Here we will add all the balls which is 10 so chances of getting red balls are 6/10(i.e., number of particular items/sums of all items)

If you try to relate all the problems which you solve with your real-life situations then terms will start looking familiar and problems will itself start giving clues for your probable solutions

Now we will try to memorize all the terms which we listen while solving probability problems. These terms can become our clues for solving our probability problems. Our terms are:

1. Experiment
2. Outcome
3. Sample space
4. Event
5. Or event
6. And Event
7. Equally likely
8. Fair
9. Unfair
10. Biased
11. Empirical

12. Complement

Your probability is mostly move around these 12 terms and we will make a relatable story so that we will have our clues ready when we have to solve probability problems

Imagine yourself as referee during the toss of a cricket game (you can imagine any game which you like the most). You have coins of all the 9 test playing nations and you always bring the coin of two playing teams on a particular day. This **experiment** is liked by all the teams as they perceive emotional value towards the coin of their country. Some feel the **outcome** will be positive if they toss their countries coin. But you explain them that their sample space is just two as two outcomes like only head or tail can come while tossing a coin. Then you explain them the rule of the **event** i.e. they can choose either batting **or** bowling **and** their team can have two extra players as replacement player in the **event** (Or /And) of any kind of serious injury. There are **equally likely** chances of keeping either one batsman, one bowler or one extra wicketkeeper and one bowler as replacement. But replacement should look fair to managing committee and if it seems **biased** or **unfair** replacement will not be allowed. So as and when required **empirical** evidence should be provided to support the replacement. Then captains of two teams **complement**ed each other and match started.

These terms can be included in any sports event situation of your liking and they are used in exact same situations where these terms are used in our probability setting.

This brings to the end of our maths chapter. I will like to suggest you to visit different math blogs, YouTube channels and websites. There are lot of free stuff available over internet and just by typing your specific problem you will definitely get either solution or clue to solve your problems. So, research methodology is the key to solve your mathematics problem and you will start playing with maths be it numbers, shapes, formulas, theorems or even statistical problems. Whoever have deep interest in maths should explore different maths quizzes, maths Olympiad and also try to visit maths museums (minimum 5 museums I know) to deepen their knowledge. This will also help you explore various career options mainly related with mathematics as major subject.

Chapter X

Physics

Physics is the branch of science which deals with matter and its relation to energy. It involves study of physical and natural phenomena around us i.e., formation of rainbow, solar eclipse, gravity, formation of shadows i.e., each and everything we notice around us. This phenomenon is taught by using principles of mechanics, optics, thermodynamics, atomic physics (quantum mechanics), electricity & magnetism and vibrations & waves. Physics the word has its origin from *physica* (Latin) and *phusika* (Greek) meaning natural things. It all started from Imhotep an architect for pyramid in 2650 BCE. 2000BC-1600 BC Babylonians collected information of plants and stars. Ancient Indians(1500BC-1000BC) explained the evolution of universe and also explained about Sun, moon and other planets. Then theory of atomism came in 475 BC (Leucippus), Atomic theory by Democritus in 460 BC. Then world understood concept of electricity and magnetism William Gilbert helped us in this in 1600. Galileo Galilei invented an astronomical telescope and contributed with Inertia, kinematics, thermoscope, parabolic trajectory of projectiles between 1564 to 1642. Theory of gravitation came in 1643 by Sir Isaac Newton. Law of conservation of energy brought by James Prescott

Joule in 1889. Thermodynamics came in to existence in 1888. Max Planck introduced formula for black body radiation in 1900. Latest development is Gravitational waves that was detected in year 2015 by Rainer Weiss, Barry Barish and Kip Thorne. Overall physics is all about very small i.e., atoms to very large i.e., universe itself. Also, the units what we use is all dependent on basic three units i.e., Length, Time and Mass. If you do not believe me please find this link https://youtu.be/X9c0MR00BzQ in which MIT Professor Walter Lewin has explained in the simplest form and you will come to know how easy this subject could be if we find our right approach. Physics in general terms depends on following 10 fields and we will understand each and every of this using daily life examples and those are

1. Atoms
2. Gravitation Force
3. Friction,
4. Electricity
5. Sound
6. Heat,
7. Force
8. Inertia,
9. Light,
10. Magnetism,

Which can be memorized by acronym AGra FrESH FILM (A, Gra, Fr, E, S, H, F, I, L, M) and you can give a lecture or speech on physics by just remembering this one acronym.

Atoms: According to definition, atoms are the smallest particle into an element can be divided without losing its(chemical) identity. So, anything and everything starts from atoms which are building blocks as we have seen to build a tower or any astronomical machine, tool or spaceship we have to start from the tiniest particle and for us that is atom. Even air around us has atoms in form of dust particle, i.e., anything in this universe has its smallest possible form is atom. So, when you are going to study atomic physics you must know the concept of atom and you will never find it difficult. Right from the simple pencil we use has Lead particles to largest Machine i.e. LHC (Large Hadron Collider) all has atoms as their building blocks.

Gravitation Force: You all know the story of Sir Isaac Newton and his curiosity leads us to discovery of Gravitational Force. As per definition gravitational force is when objects which have mass are attracted to each other the force of attraction between them is the gravitational force. So, gravity is all around us as everything has some mass and everything and everyone is attracted towards something or someone. When we make our curiosity wander freely, we also can come up with theories like Newton had brought to the world. So, keep questioning and try to find its answer till your curiosity about that subject or topic is not fully satisfied. Some of the daily examples of gravity are items heavier than air falling on the ground (e.g., if we have by mistake drop our cell phone and it will drop on ground and can be damaged) and objects lighter than air goes up like helium balloons. This will answer your other questions like aero plane flies in air though it is not lighter than air because airplane wings are shaped to make air move faster over the top of the win. When air moves faster, the pressure of the air decreases. SO, the pressure on the top of the wing is less than the pressure on the bottom of the wing. The difference in pressure creates a force on the wing that lifts the wings up into the air. Objective of this book is to awaken your curiosity about everything present around us and you will find all your answer by asking it to right person at right time. As a result of this you will never be short of self-esteem.

Friction: Do you know lighting a match stick is because of friction. Have you ever wonder why climbing on the mountain we have to apply lot of force and while coming down not much effort is needed? One of the reasons is friction. Friction from the definition is the resistance that one surface or objects encounters when moving on another. Friction is one of the most important parameters in climbing, as it decides over failure or success in climbing. Then once you have concept how friction works, then you can learn what is static and dynamic friction, coefficient of friction and its application in sports, machine manufacturing and automobile industry.

Electricity: Electricity is something which keep us charged. Yes, electricity is inside us (in our brain) and we are using electricity almost every moment right from charging our cell phones to watching TV's to even running trains and large machineries. Who do you think has invented this amazing thing for us? It's started from Benjamin Franklin who finds first association that is lightening and electricity are same and then further research took place. Thomas Edison whose contribution is most exemplary came with electrical bulb, and whole lot of other products which use electricity. George Simon Ohm who brought the concept of ratio of potential difference, current and resistance. We still evolving day by day as from Light bulb to wearable devices this journey become possible because of electricity only.

Sound: Similar to electricity we are surrounded by application of sound. Our small alarm clock to Sound Navigation ranging (SONAR) all has sound at its backbone. As per definition sound is vibrations that travel through air or another medium as an audible mechanical wave. As the definition says there has to be a medium for propagation of sound that means in the absence of air or medium there will be no sound or we cannot hear it. You would have heard about ultrasound (based on echolocation) in medical terms, our cell phones or telephone, Stringed, wind and percussion (musical) instruments all have application of sounds in them. Now whenever you heard something difficult regarding study of sound try to associate it with the daily life examples and try to understand concept and not to memorize the definition as it is and it will become your friend or start giving you clues regarding your any learning.

Heat: Heat as per definition is energy transfer from one body to another as the result of difference in temperature i.e., if two bodies are brought together temperature flows from hotter body to colder body. Some real-life examples are our sun radiates heat to keep us warm on our planet earth. In this case sun is hotter body and earth is colder body. Some more examples are Ice cube kept in to glass of water as it takes heat energy from water at normal temperature and get converted in to liquid form that is water. We mainly use four sources of heat Solar, Chemical, Electrical and Nuclear. Once you explore these four topics you will understand the applicability of heat for us. This will make you learn and understand first law of thermodynamics, followed by second law of thermodynamics and then third law of thermodynamics. Further you can research on nuclear physics and so on.

Force: Force is a push or pull upon an object resulting from object's interaction with another object. For example, if we push a door to open, we are actually applying force on it. The force applied on an object can change its state i.e., it can either bring it in to motion(bicycle) or change in shape (bean bags). How much force is needed for these changes brings us to its formula i.e.

Force= Mass X Acceleration (F= ma)

Which is also Newton's second law of motion i.e., change in velocity (acceleration = velocity/time taken) depends upon the mass of the object. This will give you concept of Unit of Force as Newton(N). As I said everything is related and by using proper association technique each and everything in a particular topic can be linked and learned easily. Our kinetics and dynamics are dependent mostly on this one concepts. Once we will see concept of Inertia everything will become clear to us. We also listen too much about vector quantity force is one of the vector quantities. Any vector quantity has magnitude (In short size and shape) and direction (movement). So, that concept is also become clear to us by its formula (F=ma).

Inertia: When your bus driver applies brake, you feel a jerk (falling forward), that jerk is inertia. By definition Inertia is the tendency of object to remain in the state of motion, or an object at rest unless acted upon by a force. We have understood concept of force and we have seen how force affect our state of motion or state of rest. This is what Newton learned and given us as his first law of motion. If you will study these patterns you also can come up with your laws that to be very soon only keep research methodology for understanding any subject. Some of the applications of the inertia is testing of airbags in automobile industry, brakes applied in running train, merry-go-round etc.

Magnetism: As we understood the concepts of atoms in every substance. Each atom has electrons in it, which carries electric currents by their spinning movements (like tops) inside the core (nucleus) of an atom. When two such substances come closer, they attract to each other. This attraction is magnetism or force applied between them is magnetic force. Now you will have a question why not all substances have magnetism. Because if the electrons are moving in the same direction, their magnetic field canceled out and we see no attraction between them. This attraction is mainly due to difference in direction of spin and number of electrons in their core. Hence some substances have strong magnetism properties and other lacks magnetism. This can also be understood from our behavior that any new place or thing attract to us i.e., we always like to explore it, due to our attraction towards it. Our earth is also a big magnet may be due to deposition of a whole lot of minerals at its core and spinning of those minerals are bringing magnetic properties which we say attraction between north pole and south pole. We heard a lot about MRI machine which is Magnetic Resonance Imaging works on the principle of Magnetism. Some more examples are car motors, railway signals and loudspeakers. Magnetism is also a type of a force which proves our concept i.e., everything is related some or the other way. Also, we have heard that opposite attracts is from the concept of magnetism only as the poles in metals i.e., North and South pole when come closer they attract to each other two south poles or two north poles actually repel.

Light: We have learned electricity and then magnetism now light is electromagnetic radiation within the portion of the electromagnetic spectrum that can be perceived by the human eye. In simple terms We need light to see. Light Comes from different sources called light sources; our main light source is the sun. Other sources include fire, stars and man-made light sources such as light-bulbs and torches. We see life in glorious color is because of the light only. Our eyes can differentiate colors because we see light in different wavelengths, for example we see rainbows in the sky in rainy season. Why rainbows have seven colors because white sun rays get refracted (changes its direction) through water droplets and gives us seven colors. As the change in direction is different so different colors of rainbows can be seen. This will lead you towards color theory and lenses. This will also make you curious towards how cameras work and how the telescope explored astronomical bodies placed at such longer distances.

This is all about learning physics in general terms and more technical terms in which physics can be broadly classified is Classical mechanics, energy and thermodynamics, electromagnetism, relativity and quantum mechanics which has been explained by Arvin Ash in his video **All Physics explained in 15 minutes** https://youtu.be/TTHazQeM8v8 on you tube. You can search for more such videos and learn from one you like.

Keep exploring different information and be curious of all the information presented to you and never stop till it completely satisfies your curiosity and you fall in love with that subject. Once you started liking any subject you will never going to get bored and your new found interest will make you master of that subject.

All the theorems and formulas can be memorized by using a technique I have explained in my podcast **Rediscover Myself** on (https://anchor.fm/chakradhar-dixit) (episodes/Memorizing-physics-formulas-using-Story-Method-to-become-Unforgettable, memorizing 10 Important Laws of Physics using association method).

Chapter XI

Chemistry

Chemistry, unlike the other sciences, sprang originally from delusion and superstition, and was at its commencement exactly on a level with magic and astrology. Even after it began to be useful to man, by furnishing him with better and more powerful medicines than the ancient physicians were acquainted with, it was long before it could shake off the trammels of alchemy, which hung upon it like a nightmare, cramping and blunting all its energies, and exposing it to the scorn and contempt of the enlightened part of mankind. The beginning of chemistry which was knows as alchemy was considered mere magic until the era of ancient Greeks. The Greek word "chemeia" first appears in the 4th century and designate to the art of metal working. It was later prefixing the article al from Arabs and art of chemistry i.e., Alchemy came in to existence. In 4th century BCE two Greek philosophers, Democritus and Leucippus brought the concept of atoms which was initially considered as four basic elements earth, air, fire and water. Till that time alchemist was using chemical reaction as ways to convert cheaper elements into gold but were not approaching chemistry in a scientific way. But in those experimentation in China, Arab kingdoms and medieval Europe alchemist do come up with some major contributions such as invention of quicksilver (Mercury) and preparation of several strong acids. But chemistry saw its real golden period not before 16th Century when Englishman Robert Boyle (1627-91) established the relationship between pressure and volume of the air through his series of experimentation. Then Joseph Priestly (1733-1804) came up with a process called Combustion, by burning carbon-containing materials in oxygen atmosphere. Then whom we regard as Father of modern chemistry Antoine Lavoisier brought the concept of Law of conservation of mass, which states that in any chemical reaction, the mass of substances that react equals the mass of the product formed. In 1803 the English schoolteacher John Dalton (1766-1844) taken Lavoisier's invention to one step further and brought the concept of Law of definite proportions which we called today as Dalton's atomic theory of matter. French Chemist Joseph Gay-Lussac (1778-1850) shown us how one volume of oxygen gas always reacted with two volumes of hydrogen gas to produce two volumes of water vapor which explains equal volumes of different gases contain equal numbers of particles which in modern day we call as **Avogadro's hypothesis** as it was actually proven chemically by Amedeo Avogadro. Then the building block of chemistry i.e., **Periodic Table** brought to us by D.I. Mendeleev (1869) in Russia which had seen lots of revision as per discovery of different elements and metals. Organic chemistry came in to existence not before the 19th century and it was because of Friedrich Wohler's synthesis of Urea (1858). Then different concepts like valency by Edward Frankland (1852), benzene ring structure by F.A. Kekule (1858) started to become significant and organic chemistry had become relevant for scholars. Thermodynamic equilibrium pioneered by Henri Louis Le Chatelier (1850-1936). Alfred Nobel had given dynamite to the world in and also established international awards for achievement in the field of science (Chemistry, physics and medicine). Cathode rays brought the concept of electrons existence and this was the achieved by sir J J Thomson (1856-1940). Theory of valence was the contributed by A. Werner (1866-1919) and Madam Curie (Marie Curie) shown the radioactivity presence and its benefits to the world and also won Nobel prize for her extraordinary achievement. We understood the concept of neutron by Sir James Chadwick (1932) and First controlled fission reaction was done by Enrico Fermi between 1939 to 1942. We understood the structure of DNA Molecule in 1953 with the effort of FHC Crick and James D Watson. Afterwards many continuous researches and inventions in the field of elements of periodic table (e.g., Plutonium and Seaborgium) and chemistry of life like vitamin B12 and Human chemistry like DNA, RNA and Amino acids etc. had come in to existence. Even now the continual inventions and researches are taking place to make human life easier. Refer http://www.arvindguptatoys.com/ this website and you will come to know how one can enjoy learning chemistry(science) in fun and easy way that to be without much investment. Lot of people are doing lot of good things we need to reach them or help them the way we canso that this world can be more in equilibrium for all.

On this note let us see some concepts of chemistry like how to memorize the whole periodic table and please find below the old periodic table with fairly new approach and you will understand how one can change the difficult concepts in small bitesize easy chunks and learn and memorize so that it can come handy in later stages of life. In the below table I have categorized the elements as per their uses and it can help you in remember the list of elements by the association technique and will also help you remember two points one is their atomic number and another their uses so you can describe those elements and also as you know their uses you can learn what are the applications of particular element. These applications will make you understand each and every element in detail and in later stage of life whenever you will see any balloon you will have the idea that it is making use of Helium or mobile batteries will give you clue for lithium and you can relate with the information you have learned in past. These relations are our keys to recall any information and more we make associations more chances of recalling any information will be there. I have done it for 15 element and you can do it for all 118 elements and keep it handy for your spaced repetition practice.

Sr. No.	Elements	Uses
1	Hydrogen	
2	Helium	

3	Lithium	
4	Beryllium	
5	Boron	
6	Carbon	
7	Nitrogen	
8	Oxygen	
9	Fluorine	
10	Neon	

11	Sodium	
12	Magnesium	
13	Aluminum	
14	Silicon	
15	Phosphorus	

It has been said that learning periodic table is learning half the chemistry and I have given you the trick to remember half the chemistry and your stress level should be lowest by now as fear is not good for grades. If you are still skeptical that how you are going to memorize the whole periodic table break the elements in chunk of 15 elements and make 8 stories around them using their applications and try to memorize each story almost every day and in less than 15 days all the elements will become part of long-term memory. Also keep a revision schedule as per spaced repetition cycle and it will never ever go out of your brain. One or two elements which initially you will have issues in recalling will also be recalled easily after second or third space repetition cycle. Key to memorizing is always make stories which are vivid and which can be seen clearly and take my word you will never ever forget these elements again in your life.

Now I will tell you the ways to remember the chemical formulas and it is also going to be exercise filled with lot of fun. I will tell you the chemical reaction for photosynthesis with the help of a story.

$$6CO_2 + 6H_2O \xrightarrow{\text{Sunlight energy}} C_6H_{12}O_6 + 6O_2$$

Where: CO_2 = carbon dioxide
H_2O = water
Light energy is required
$C_6H_{12}O_6$ = glucose
O_2 = oxygen

Our story for this could be an Arab is wearing diamond (Carbon) shoe (6) has come in to a hospital to meet his wife Noah (2) who is in ICU and doctors are kept her on life support system (Oxygen). But she sees her husband she gets up and ask for about Fertilizer business and Arab shows his shoes (6) i.e., he had converted everything in diamond and made shoes out of it. She again started breathing heavily and doctor had to keep her on Life support system (O) again. Doctors advised Arab that if he wants her alive then he should give her fertilizer business back to her. Arabs then goes in sunlight and after getting some energy he comes back with Diamond Shoes(C6), Fertilizer Tin (H-Fertilizer ,12-Tin) and Life support system payment slip which is kept in shoes(O6). His wife becomes happy and doctors turn the knob off by their shoes (6) and take her out off from Life support. Arab thank the doctors and took Noah (2) home.

Here the characters (Actors) of your stories are:

Carbon- Diamond (From above table)

Hydrogen- Fertilizers (From above table)

O- Life Support System (From above table)

6- Shoes (From Phonetic Number Sound System)

2- Noah (From Phonetic Number Sound System)

12-Tin (From Phonetic Number Sound System)

Sunlight and Energy is also taken in to context of story

It is the story which I have used to memorize this whole chemical reaction. You can use your own associations and memorize even more complex formulas also. Your stories could be more memorable for you as you will be doing all the associations and linking by yourself.

These are certain ways by which you can memorize and recall the complex information in chemistry. If you want to memorize all the laws and timelines you can go for keyword method and by making stories around those points you can memorize all the complex information which you want to memorize and recall.

Chapter XII

Biology

The word biology is derived from the Greek words "*bios*" meaning life and "*logos*" meaning /study/ and is defined as the science of life and living organisms. An organism is a living entity consisting of one cell e.g., bacteria, or several cells e.g., animals, plants and fungi. Though history of study of Biology not known clearly but it can be traced back to Assyrian and Babylonian era as per "Britannica Online Encyclopedia" i.e., they were using as veterinary medicine. Carolus Linnaeus on Jan 1, 1740 comes first with binomial nomenclature. Then three German scientist's Matthias Schleiden, Theodor Schwann and Rudolph Virchow brought the concept of Cell Theory in 1855. Then Darwinism came in to existence i.e. Survival of the fittest i.e., Charles Darwin came with his evolution theory of natural selection in the year 1859. In April 1953 Double helix model of DNA was proposed by James Watson and Francis Crick. Karry Mullis got the Nobel prize for his PCR (Polymerase Chain Reaction) in December 1983.

Biology is divided in to three broad categories i.e., Botany, Zoology and Microbiology. If we expand our view some of the main branches of biology are as many and can be classified as follows:

1.Taxonomy: It is the science of identification, nomenclature and classification of organisms. Carolus Linnaeus believed to be father of Taxonomic system i.e., who had given the concept of classification of plants and animals.

2. **Morphology**: It is the study of external form, size, shape, color, structure and relative position of various living organ of living beings. Wilheim Hofmeister was believed to be one who had brought the concept of Plant Morphology in 1853.

3. **Anatomy**: It is the study of internal structure which can be observed with unaided eye after dissection. Andreas Vesalius was a Belgian born anatomist and physician, is considered the father of modern anatomy and he explained this as Fabric of human body in his book *Fabrica* written in Latin.

4. **Histology**: It is the study of tissue organization and structure as observed through light microscope. Marcello Malpighi described a series of microscopic structures never seen until then; for instance, was the first scientist to observe the capillaries. He is considered to be father of Histology and this field is further seen lot of inventions credit goes to Robert

Hook (1635-1703), Max Schultze (1861) and Leeuwenhoek (1700).

5. **Cytology**: It is the study of form and structure of cells including the behavior of nucleus and other organelles. George N Papanicolaou who had brought the concept of Pap test (procedure to test for cervical cancer in women) and he actually brought in front the work of Robert Hooke and put it in the form of Cytology for betterment of understanding the analysis and cure of cervical cancer.

6. **Cell Biology**: It is the study of morphological, organizational, biochemical, physiological, genetic, developmental, pathological and evolutionary aspects of cell and its components. It had been brought by Robert Hooke in 1665 and further concepts of Bacteria and protozoa came in to existence by Van Leeuwenhoek. This link is your guide to understand it in more detail ([(1) The wacky history of cell theory - Lauren Royal-Woods - YouTube](#))

7. **Molecular Biology**: It is the study of the nature, physicochemical organization, synthesis working and interaction of bio-molecules that bring about and control various activities of the protoplasm. The field of molecular biology arose from the convergence of work by geneticists, physicists, and structural chemists on a common problem: the nature of inheritance. In the early twentieth century, although the nascent field of genetics was guided by Mendel's

laws of segregation and independent assortment, the actual mechanisms of gene reproduction, mutation and expression remained unknown. Thomas Hunt Morgan and his colleagues utilized the fruit fly, *Drosophila melanogaster*, as a model organism to study the relationship between the gene and the chromosomes in the hereditary process (Morgan 1926; discussed in Darden 1991; Darden and Maull 1977; Kohler 1994; Roll-Hanson 1978; Wimsatt 1992). A former student of Morgan's, Hermann J. Muller, recognized the "gene as a basis of life", and so set out to investigate its structure (Muller 1926). Muller discovered the mutagenic effect of x-rays on *Drosophila*, and utilized this phenomenon as a tool to explore the size and nature of the gene (Carlson 1966, 1971, 1981, 2011; Crow 1992; Muller 1927). But despite the power of mutagenesis, Muller recognized that, as a geneticist, he was limited in the extent to which he could explicate the more fundamental properties of genes and their actions.

8. Physiology: It is the study of different types of body functions and processes. Claude Bernard is often referred to as the father of modern experimental physiology. He was born 200 years ago, on 13 July 1813, in France. He has given the concept of Homeostasis (any self-regulating process by which biological systems tend to maintain stability while adjusting to conditions that are optimal for survival).

9. **Embryology**: It is the study of fertilization, growth, division and differentiation of the zygote into embryo or early development of living beings before the attainment of structure and size of the offspring. Karl Ernst von Baer (1827) was an Estonian professor studying embryos and development when he made a discovery that laid the foundation for modern comparative embryology.

10. **Ecology:** It is the study of living organisms is relation to another organism and their environment. Ecology was originally defined in the mid-19th century, when biology was a vastly different discipline than it is today. The original definition is from Ernst Haeckel, who defined ecology as the study of the relationship of organisms with their environment.

11. **Genetics:** It is the study of inheritance of characters or heredity and variations. Heredity is the study of expression and transmission of traits from parents to offspring. The way in which traits are passed from one generation to the next-and sometimes skip generations-was first explained by Gregor Mendel (1862). By experimenting with pea plant breeding, Mendel developed three principles of inheritance that described the transmission of genetic traits, before anyone knew genes existed.

Chromosome theory, genetic linkage Mitosis, Meiosis and Inheritance are some of the concepts which started coming out for better understanding of this branch of Biology.

12. **Eugenics**: It is the science which deals with factors related to improvement or impairment of race, especially that of human beings. In 1883, Sir Francis Galton, a respected British scholar and cousin of Charles Darwin, first used the term eugenics, meaning "well-born." Galton believed that the human race could help direct its future by selectively breeding individuals who have "desired" traits. This idea was based on Galton's study of upper-class Britain. Following these studies, Galton concluded that an elite position in society was due to a good genetic makeup. While Galton's plans to improve the human race through selective breeding never came to fruition in Britain, they eventually took sinister turns in other countries.

13. **Evolution**: It studies the origin of life as well as new types of organism from the previous ones by modifications involving genetic changes and adaptations. If we believe to Harvard study then, Charles Darwin is commonly cited as the person who "discovered" evolution. But, the historical record shows that roughly seventy different individuals published work on the topic of evolution between 1748 and 1859, the year that Darwin published On the Origin of Species. You can follow this link to understand the concept of

evolution better- Who Discovered Evolution? - YouTube

14. **Paleontology:** It deals with the study of fossils or remains and impressions of past organisms present in the rocks of different ages. Georges Cuvier & William Smith (1800), considered the pioneers of paleontology, found that rock layers in different areas could be compared and matched on the basis of their fossils.

15. **Exobiology:** It is the branch of scientific inquiry dealing with the possibility of life in the outer space. The term exobiology was coined by molecular biologist and Nobel Prize winner Joshua Lederberg in 1960. and the field grew significantly with space exploration, especially the Viking landers on Mars. Exobiology draws largely from four disciplines: planetary science, planetary systems science, origins of life studies, and the Search for Extraterrestrial Intelligence (SETI). The field has been invigorated by claims of fossil life in an ancient Mars rock, the discovery of a possible ocean on the Jovian moon Europa, extrasolar planets around sun-like stars, life in extreme environments on Earth, and complex organic molecules in interstellar molecular clouds. Life itself, however, has not yet been found beyond Earth.

16. **Virology:** It is the study of viruses in all their aspects. Martinus Beijerinck is often called the Father of Virology. Beijerinck published his findings, calling

his discovery a *'contagium vivum fluidum'*, or virus. The name virus was coined from the Latin word meaning slimy liquid or poison. It was originally used to described any infectious agent, including the agent of tobacco mosaic disease, tobacco mosaic virus. In the early years of discovery, viruses were referred to as filterable agents. Only later was the term virus restricted to filterable agents that require a living host for propagation.

Why am I telling you all the history because stories(history) actually make the subject learning interesting and you can grasp it better. I have shared lot of video links from ted.com and other educational sites and you can clearly see the pattern of story-telling, use of animations and concept building as the mode of delivery. So, develop a habit of making stories around your difficult concepts and dig deeper it will automatically start becoming interesting for you. Also, there are minimum 16 branches of biology and not just two or three as we all know and there is lot of scope in each and every field. What is required by us to study and explore the field which we like the most and we can bring the significant changes in the world. Now let us turn to the field of communication and understand widely used languages and how we can learn them to get benefits of actual globalization and can move around the world with open mind and mix with different cultures.

Chapter XIII

Foreign Languages

There is a database of 7117 spoken languages across the world and when you have been asked how many languages you can speak, you can hardly tell that two or maximum three, if you are not polyglots. If we will not understand the language of the person with whom we have to start a conversation it is very difficult to understand each other. So there has to be an effective tool to learn a new language or at least convert it something which we can understand and able to reply and also make the person understand your feelings through your language. English, Mandarin and Hindi are the three major language of the world spoken by 3000 Million i.e., by approximately by 42% World Population. But there are a staggering figure of 2926 languages, which are endangered (i.e., also approx. 41% of total languages) today i.e., almost at the verge of extinction. So, if we want to understand the complete world history, understanding different languages and hence different cultures are must. Also learning any new language is beneficial for our proper brain development as it enhances our thinking skills and memory recall abilities. As we have understood why it is important to learn any new language let us now understand the history and evolution of languages for better learning experiences. If I have to tell you the history of each and every language at least 100 such books can also not cover the whole history part itself. So, I will tell you the dialect part and their distribution to the whole world due to human migration.

It's difficult to decide the origin of the language as we do not have any formal tool to do so. But there are people who did research and found out for us that historical linguists compare large numbers of words in different languages, and they found out that tea is called either "te" or "cha" due to its origin and trade from China. We can always find some similarities between Greek and English words like pater in Greek is Father in English, like wise podos is foot. The common ancestor of English, Latin, Greek, Russian, Gaelic, Hindi, and many other languages spoken in Europe and India is known as Proto-Indo-European, whereas the more recent common ancestor of just English, German, Dutch, Norwegian and the other Germanic languages is known as Proto-Germanic. These video links - (1) Verner's Law, Part 1 of 3 - YouTube, (1) Verner's Law, Part 2 of 3 - YouTube, (1) Verner's Law Part 3 of 3b - YouTube is recommended to understand the origin of various languages and this will also help you understand the whole world is united and can be traced back to common origin and also learning foreign languages may not be that difficult as we think. As we have got the idea how the languages are related to each other now let us come back to memory methods approach to learning any language.

The very first thing we learn about language is its vocabulary and that can be learn using "Keyword Method" or we can call it word substitution method also. This the same method which we have used to memorize long speeches in earlier section, but here we will make picture association using the words in new language and in our native language.

Suppose you have to learn French, and how do we say "thank you" in French, we say Merci.
So here we can make a picture or story like "Your friend Mary see and tank and say thank you".
Likewise, for words like "Bonjour" is "good morning" is French, so here you can make a story like "There is no light in the camp and you are lighting a bon fire and your French friend thought it's sun light and is saying good morning to you". You can make such weird stories and record it in your long-term memory. This will be your clue for recalling the words when you will test it in actual language speaking scenario.

Some more tips could be to prepare 5 words of any new language and associate with it with pictures in your native language and within a year you will have more than 1500 words in your vocabulary and also keep recalling it using spaced repetition method i.e.(1day-1week-1month-3month-1year) and you will never forget those words again. But revision is must. I will share some more tips also, like learning languages using mind map®. In this you can take one word and try to mark all the items and sentences related to it in single mind map. This will cover everything related to the base word. For example, if we take example of Cherry fruit and mark all the things related to cherry in single mind map®, this will help you understand how many products can be made using that material and how local people use that word in their native language. This can be illustrated as given in the mind map below:

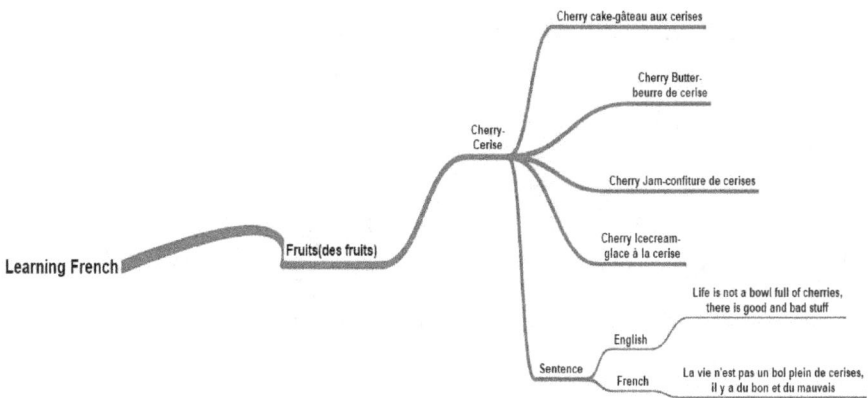

Similarly, we can do for other fruits, spices and even the materials in daily use. By using mind maps® and making flash cards out of these words we can prepare many keywords and slowly and steadily we can learn that language.

Being a student it's too much an ask but if you have an idea of podcast then you should listen podcast in the language which you want to learn and you will slowly start getting ideas about words and pronunciation in that language and learning that language will become easier for you.

I have explained this for French, but the same thing can be done for German, Spanish, Mandarin, Hindi, Arabic or any other language which you want to learn. Also, I am sure that once you will start taking interest in local culture and understanding dialects, you can easily relate with that language and your AIM (Attention, Interest and Motivation) will bring you closer to learn that language.

I have mentioned flashcards which you can make online using Anki App and also for Mind Maps many free softwares are available but making it analog way (by hand) is advisable as it will bring your kinesthetic memory in play and learning will become fun. On this note let us come to learning history of our subject

History. Before that here are some the links which will motivate you to learn new languages and will tell you, how learning foreign languages are actually good for your brain.

1. [(1) How to learn any language in six months | Chris Lonsdale | TEDxLingnanUniversity - YouTube](#)
2. [(1) 5 techniques to speak any language | Sid Efromovich | TEDxUpperEastSide - YouTube](#)
3. [(1) Learning a language? Speak it like you're playing a video game | Marianna Pascal | TEDxPenangRoad - YouTube](#)
4. [(1) Creating bilingual minds | Naja Ferjan Ramirez | TEDxLjubljana - YouTube](#)

Ok, I think this much motivation is enough for learning foreign languages. I know this will lead you to explore a lot of free learning stuff over internet by yourself.

Chapter XIV

History

The word history has derived from the Greek word historia that means the act of seeking knowledge or enquiry. So, history actually enquires about the events happened in past. Herodotus is considered to be "father of History", as he was the leading source of historical information of Greek as well as Western Asia and Egypt between 550 to 479 BCE. This one link is your complete guide of world history- (1) Timeline of World History | Major Time Periods & Ages - YouTube. My aim of writing this book is to make you understand that each and every subject which we study, have some meaning and it is important to keep all the information in our brains for rest of our lives. That's why I am sharing so much of links, videos and images so that your mental movie can keep giving you clues for any information which you have to recall.
I will help you with the dates i.e., how to memorize the important dates from history by using phonetic number sound method.

Some important dates in the world history are: -

1. **3500 B.C. – Invention of wheel**
   It can be memorized by using 35- Mole and 00- See as images from our phonetic number sound method and our story can be **Mole sees a wheel** and start running towards British Columbia (B.C.)

2. **486 B.C. – Birth of Buddha**
   It can be memorized by using 4- Ear and 86- Fish as images from our Phonetic number sound method and our story can be Fish listens using human like Ear, the crying of cute Buddha as a child.

3. **1176 A.D. – American declaration of Independence**
   It can be memorized by using 17- Duck and 76- Cash i.e., Duck is getting a lot of Cash as it's become independent business owner.

4. **1917 A.D. – Russian Revolution**
   It can be memorized by using 19- Tub and 17- Duck i.e., a Tub is full of Duck from various types and they are on a strike and they are holding a Russian flag. They are demanding the proper sanitization of Tub water.

5. **1947 A.D. – India become Independent from British rule**
   It can be memorized by using 19- Tub and 47- Rack i.e., Tubs and Racks filled with flowers are brought at India gate to celebrate Independence Day of India from British rule.

6. **1949 A.D. – Communist China Founded**
   It can be memorized by using 19- Tub and 49- Rope i.e., Many Tubs are tied using a common Rope at opening of Great wall of China for public.

This way you can remember as many dates as you want by making simple stories, but as I have told you earlier, make totally unrelated stories, as our mind needs trigger to give back any information, which we want, so colorful and strange stories, could be the interesting concepts for our brain and keeping such images are easy and recalling them can be faster.

Let us understand something about our mother earth that is Geography now.

Chapter XV

Geography

The word geography can be broken down into the two basic elements of "GEO" and "GRAPHY." *Geo* comes from the Greek word for Earth (the word *Gaea,* also meaning earth, derives from the Greek as well). The "graphy" part comes from the Greek word *graphein*, which is literally to write about something. Records can be found of as old as 300 million years of Pangaea or Pangea supercontinent that was existed during the Paleozoic and early Mesozoic eras. It was the only continent which was there on the earth but then it started breaking apart and seven different continents came in to existence. This was the time when Dinosaurs were on the earth. Studies are still under progress to find out exact details. Then next known records show that Three and a half thousand years ago, the tiny Aegean island of Thera was devastated by one of the worst natural disasters since the Ice Age - a huge volcanic eruption. This cataclysm happened 100km from the island of Crete, the home of the thriving Minoan civilization. Pytheas, (300 BC) was a navigator, geographer, astronomer, and the first Greek to visit and describe the British Isles and the Atlantic coast of Europe. Though his principal work, On the Ocean, is lost, something is known of his ventures through the Greek historian Polybius (200BCE–118 BCE). Eratosthenes of Cyrene, (276 BCE—194BCE) was a Greek scientific writer, astronomer, and poet, who made the first measurement of the size of Earth for which any details are known. He did the measurement of size of earth by noting down angles of shadows in two cities on the Summer Solstice (a solstice happens when the sun's zenith is at its furthest point from the equator. On the June solstice, it reaches its northernmost point and the Earth's North Pole tilts directly towards the sun, at about 23.4 degrees.), and by performing the right calculations using his knowledge of Geometry. His calculation was remarkably accurate. He also created the first map of the world incorporating parallels and meridians, based on the available geographical knowledge of the era. Hipparchus (190 BCE- 120 BCE) had very accurately cataloged over 1,000 stars and invented the mathematical science of trigonometry. Claudius Ptolemy (90AD-168AD) was a Greek mathematician, astronomer and geographer. Much of medieval astronomy and geography were built on his ideas: his world map, published as part of his treatise Geography in the 2nd century, was the first to use longitudinal and latitudinal lines. Piri Reis (1465-1553) was a sixteenth-century Ottoman Admiral famous for his maps and charts collected in his Kitab-ı Bahriye (Book of Navigation), a book which contains detailed information on navigation as well as extremely accurate charts describing the important ports and cities of the Mediterranean Sea. Antonio Pigafetta, a Venetian scholar and explorer was one of the 18 men who returned to Spain in 1522, under the command of Juan Sebastián Elcano, out of the approximately 240 who set out three years earlier. These men completed the first circumnavigation of the world. Abraham Ortelius (1527-1598) is a key figure in the history of human knowledge. He is known as the inventor of the atlas - a book bringing maps together in one format and with the same display - and was the first person to discover

continental drift. British navigator James Cook in 1779, charted New Zealand and Australia's Great Barrier Reef on his ship HMB Endeavour.
Pedology(is a discipline within soil science which focuses on understanding and characterizing soil formation, evolution, and the theoretical frameworks through which we understand a soil body(s), often in the context of the natural environment), general use of this term began in Russia following development of a new approach to the study of soils by Dokuchaiev and his students in the late 1870s and early 1880s. William Morris Davis, (1850- 1934), U.S. geographer, geologist, and meteorologist who founded the science of **Geomorphology**, the study of landforms. Then different geographical societies started coming up like, like the Royal Dutch Geographical society was founded in 1873 in line of similar groups in other major European countries, such as France (1821: Société de Géographie de Paris), England (1830: Royal Geographical Society), Germany (Berlin, 1828, 1836 Frankfurt, Munich, 1869, Bremen 1870 Hamburg, 1873, Leipzig 1861), and Russia (St. Petersburg] 1845). So, study of geography as a subject started peaking up. The term geopolitics was first used by the Swedish political scientist Rudolf Kjellén (1864-1922), a student of the German geographer Friedrich Ratzel (1844-1904). His ideas formed what is known today as the perspective of "classical" **Geopolitics**. Notable steps towards the humanistic tradition in geography were taken by Paul Vidal de la Blache (1845-1918) and his students in the late 19th and early 20th centuries. They introduced a *point de vue* to geography (Berdoulay 1995: 184). This is what Vidal himself calls 'terrestrial unity' (l'unité terrestre). Due to his effort the term '**Human Geography**" was came in to existence. In 1904, Sir

Halford Mackinder published the Heartland theory (any political power based in the heart of Eurasia could gain strength to eventually dominate the world.). The theory proposed that whoever controls Eastern Europe controls the Heartland. It also supported the concept of world dominance. Explanation - A more revised version explains that whoever controls the heartland, controls the world island. In the 1920s, Carl Sauer became influential in urban geography as he motivated geographers to study a city's population and economic aspects with regard to its physical location. Walter Christaller developed his "Central Place Theory" in the 1930s. This theory is based on his idea that settlements only existed to function as "central places" to provide services for the surrounding area.

This theory is part of the study of urbanization, taking into account the importance of supply and demand. This is the broad view of Geography as a subject and how our earth and our studies of earth changed over time. Just by following the history of Geography you can understand the need for various developments and how useful those ideas and inventions are for human being.

By making you understand I am trying solve your query of why we study any subject and how vast a subject can be if we start exploring from scratch. One final link to understand the world and its upcoming challenges with geographical point of view, so that you can understand the present situation better and plan your futures accordingly- Maps that show us who we are (not just where we are) | Danny Dorling - YouTube.

Different classifications and answers can be memorized using the memory techniques already explained in earlier chapters. With this we have come to last section of this book and hope this book will help you understand your curriculum better and will also give you resources for your out of the box thinking and will let you clear your mind for any limiting beliefs so that you can become freely ask any questions, without fear and get the desired knowledge. As knowledge is the power to change the world in a better place and by knowing history you can understand everyone is connected some or the other way and there is division just on the basis of movement of population from one place to other, otherwise everybody is same and everyone is associated and connected to common link. On this note, thank you very much for finishing this book and I will come with some more knowledge sharing part as I will explore the ocean of knowledge.

Be Fearless and Be Unforgettable.

I want to check with you, if you want a chance to win three free eBooks and an Audio Book of "Unleash Your Memory".

Hey, did you know that your brain has storage capacity of 2.5 petabytes, which is enough to run for HD quality videos for 300 years.

I know it sounds crazy because, we are so used to hearing about memory recall problems, and while that is important, I'm going to share three secrets with you today that are going to revolutionize the way you can train your memory to keep 3x times more data in your brain and can recall it whenever you want it.

I am Chakradhar Dixit, and I am a Memory Coach who has helped more than 1000 people to get their 10x memory power.

So, what is the secret of getting minimum 3x Memory power

It's just 3 secrets

It's to change your eating habits, do some mental exercise and try using memory tricks or training to recall all your valuable information.

So, I'm going to share with you three secrets about increasing your memory recall power and how you can become unforgettable.

And I know what you're probably thinking if is it going to be that easy why you have not heard it before?

But again, I'm going to show you a great way you can use these three secrets and get amazing results.

So, the first secret is to include Avocado, Blueberries, Broccoli, Plant oil, Eggs, Green Leafy Vegetables, Salmon Fish, Turmeric (Curcumin), Walnuts, Dark chocolates and enough water.

The big idea here is these brain foods will keep your brain forever young and will give all the nutrients it needs to function at the optimum level.

This is important because our brain has scientifically proven plasticity that is it can be improved at any stage if we care for it and given the essential nutrients.

The second secret is mental exercise i.e., using your non dominant hand for brushing your teeth, hand-writing your scripts, playing chess and sudoku and reading books daily.

The main thing to understand here is that mental exercises are not that difficult as the physical exercises but gives multifold results in building your smarter and brighter brains.

This means by doing mental exercises you build your brain muscles to give you better ideas and helps in better functioning of your brain. Researches have proved that mental challenging games even keep you away from Alzheimer's' to certain extent.

The third secret is memory training or memory tricks for recalling complex information in shortest possible time. The three widely used memory tricks are association technique, memory palace and story method.

The main thing to understand here is these memory techniques are known to us from our course books but we never use it in our daily lives. To tell you an example we learn our ABCD by means of pictures i.e. A for Apple, B for Bat but we do not associate our complex information to anything which we already know and hence we are not able to recall it when we need it.

This is the key because memory training is your answer for all the questions and issues related to your memory recall such why I forgot certain piece of information or things or the person's name whom I met earlier today? Why I am not keeping the information in my mind which I have read last week or how to speak confidently in public speaking event without using cheat sheets?

Now, I know what you're thinking.

It cannot be that simple or how I can be so confident about this and if it is so simple why so many people are afraid of public speaking and even so afraid that they will prefer dying that to deliver a speech on stage?

Well, the tricky thing is that we all are so busy in finding solutions to our problems in complex and most tedious manner that we never realize that we are the owner of the fastest super computer and it needs a very little care and practice to run it on optimal level and get maximum benefits out of it.

So, I've created this Memory Unleashing Blueprint by using this you can become unforgettable in no time.

It's so effective that I have unlocked my potential of being an author, a podcaster and even a memory coach within 6 months of time and my students are getting amazed by their memory power as it shows results in as early as within a week of taking this course.

And so, what this is going to do is it will make you remember all the names of people you meet, learn any foreign language, use your brain's full potential and inculcate the habits which will bring the best version of yours hidden inside you. You can do whatever you have dreamt of for yourself.

We're always here to help and I can't tell you all the amazing results we've gotten for students, corporate employees and seniors just like you who are struggling with their untrained memories.

We also have 1-2-1 Coaching, but today I'm talking about Memory Unleashing Blueprint, which is amazing.

I can't wait to see you on the inside. Here is your link:

Lifechanging Links (Free Gifts Inside)

For all: Memory Unleashing Blueprint | Udemy

For Entrepreneurs:
https://www.funnelchallenge.com?cf_affiliate_id=2802528&affiliate_id=2802528

---

[i] Reference-https://www.gwern.net/Spaced-repetition

[ii] (<a href='https://www.freepik.com/vectors/background'>Background vector created by macrovector - www.freepik.com</a> <a href='https://www.freepik.com/vectors/background'>Background vector created by macrovector - www.freepik.com</a>)

[iii] (https://www.politicalscienceview.com/what-are-the-different-types-of-governments/)

[iv] (https://www.edutopia.org/discussion/your-exams-do-not-determine-your-worth)

www.ingramcontent.com/pod-product-compliance
Lightning Source LLC
Chambersburg PA
CBHW060841220526
45466CB00003B/1192